Stand Your Ground Forecasting

Mastering the qualitative and soft skills for forecasters and business analysts

Don Ake

Also By Don Ake

Just Make Me A Sammich

Will There Be Free Appetizers?

Turkey Terror At My Door!

Deep Heavy Stuff

Copyright 2024 Don Ake & Wojelay Publishing

All Rights Reserved

Table of Contents

Introduction	1
Chapter 1 – Hey, We Need A Forecast for This	4
Chapter 2 – The Relentless Pursuit of Accuracy	17
Chapter 3 – The Heart of a Forecaster	27
Chapter 4 – Develop A Process – Trust The Process	48
Chapter 5 – Becoming A Master Forecaster Who Stands Your Ground	60
Chapter 6 – Setting The Foundation For A Winning Process	68
Chapter 7 – Fundamental Forecasting Issues	72
Chapter 8 – Establishing An Accuracy Standard	81
Chapter 9 – Building Your Information Network	89
Chapter 10 – Building Your Data Collection	101
Chapter 11 – Making And Documenting Forecast Assumptions	109
Chapter 12 – Getting To Your Forecast Numbers	123
Chapter 13 – Measuring And Reporting Forecast Accuracy	142
Chapter 14 – Refining Your Process – Tightening The Standard	157

Chapter 15 - Defending Your Forecast	172
Chapter 16 – "Meeting" Expectations – The Forecast Meeting	195
Chapter 17 – Putting It All Together – The Monthly Forecasting Process	207
Chapter 18 – Miscellaneous Forecasting Issues	218
Chapter 19 – Demand Vs. Financial Forecasts.	233
Chapter 20 – How To Estimate Anything	245
Chapter 21 – Dealing With Caustic Company Politics and Other Corporate Crap	254
Chapter 22 - Becoming A Master Forecaster Who Stands Your Ground - Part 1 Control & Competence	274
Chapter 23 – Becoming A Master Forecaster Who Stands Your Ground - Part 2 Credibility & Confidence	281
Chapter 24 – Forecast Well And Stand Your Ground	295

Introduction

The idea for this book was hatched sometime during a company meeting in 2020. My coworker Avery Vise was amused by one of my forecasting philosophies, and I blurted out, "Yes, that's going to be a chapter when I write my book on forecasting." The meeting moved on to another topic, but those words hung in the air and captured my attention for the next five minutes. This was one of those ideas that was so obvious that you can't believe you didn't think of it earlier. At that time, I had already written three books (Short-essay humor books), and I had over 17 years of forecasting experience. So, it was a major no-brainer, and it became my top priority after I retired in 2022.

In a sense, this is a giant brain dump of everything I learned about forecasting and analytical marketing over my career. In addition to revealing the technical process that I developed, I spend a lot of time discussing the soft skills needed to navigate the oftentimes treacherous waters of the corporate world, skills that do not come naturally to highly analytical people like myself and where I must say, learned the hard way. It is all presented here to help you be better at your job and survive in a perilous political environment. One title considered for the book was "Swimming With The Forecasting Sharks", for obvious reasons.

I share what I learned over my 40-year career, organized and expounded upon for your benefit. I want you to prosper from reading about my successes and failures. The book is a treasure trove of ideas that will not only enhance your job performance but also catapult your career to new heights. It is not a how-to book because every process and situation are different. I detail my process, but I want you to take the principles and ideas and infuse them to create/improve your own process.

I took my job as a forecaster seriously. When I got knocked down, I not only got back up, but devised strategies so as to not get knocked down that way again. A forecaster is like a boxer – you must bob, weave, dodge, block the uppercut, and roll with the punches. I was committed to forecasting excellence. I wasn't always the most accurate forecaster in my industry – but I always considered myself the best.

Even though this book is primarily about business forecasting, any analytical business person will benefit from the concepts and tactics detailed here. Business analysts, market researchers, financial analysts, stock pickers, pricing people, accountants, etc., will be able to apply much of the content to their jobs. In addition, non-business forecasters and frankly anyone in the business world, can also learn some soft skills and enhance their business communication abilities. Since many of these concepts are universal, I also expect the book to have some international interest.

If you pick up one concept, strategy, trick, maneuver, etc., that you can use repeatedly over your career, then this book is worth many times the price you paid – or your company, if you expensed it. And regardless, you will love the colorful but relevant stories from my career, strategically presented throughout.

Now, roll up your sleeves. It's time to get to work and learn to Stand Your Ground

Chapter 1
Hey, We Need a Forecast for This

Stand Your Ground

Forecasting and business analysis are tough jobs. Your numbers and predictions are always scrutinized and challenged. At any moment, you can become the company punching bag, whipping boy, or dunce. I want you to produce the best forecasts/predictions possible. Then, be able to defend your numbers – and stand your ground.

This is a different type of forecasting book because it deals with the soft skills and qualitative aspects of the task. It has been said that forecasting is both an art and a science. This book is all about the art part, and I want to make you an artist!

Why We Are Here

This book will benefit anyone who uses data, information, and analysis to make forecasts, predictions, and recommendations in a business/organizational setting. This includes forecasters, market analysts, business analysts, economists, and marketing researchers. I will use the term "Business Predictor" to describe the functions other than forecasting.

It is written from a forecaster's perspective because that is what I know best. However, the principles presented here are pertinent to many business positions, futurists, and other non-business predictors. Therefore, please adapt them to your specific situation.

The book will turn you into a Master Forecaster, who can stand your ground. Someone who consistently produces accurate forecasts and is an expert in your field. Or a Superior Business Predictor, making many more good calls than bad ones.

In addition, I want to help you survive and flourish in your job. Forecasting is a risky profession, and the processes, tactics, and strategies espoused here will enable you to handle the corporate political pressures you face.

In discussing the concept of this book with colleagues outside of the forecasting realm, I discovered the content on managing up and surviving a challenging corporate political environment, is of interest to all businesspeople. So, if there is only one concept presented that helps you in your career, this book was a worthwhile investment.

The Beginning of the Journey

Few people aspire to be forecasters, although forecasting did intrigue me at an early age. I had to write a report in 7th grade on a future career and chose meteorology. I went to the airport and interviewed the meteorologist. The taped interview featured the astute question: "There is a lot of complicated equipment in this place. Is there any chance of it blowing up?" And even though that meteorologist was safe in his environment, as a forecaster, there are always the pitfalls of something blowing up.

I was initially a college statistics major. My school had a top-level engineering program and used the Calculus II course to thin the herd. I averaged a "D" grade in the second half of Calculus I, so there was no chance of getting through the subsequent class. I may be unable to handle derivatives and integrals; what I do have is the innate ability to look at a chart or graph and see things others miss. I ended up changing my major to marketing. My master's thesis was "Developing a Concept Testing Model for Industrial New Products", so I was destined to end up on the analytical side of marketing.

Many people do aspire to be business predictors. College programs/majors for business analysis/analytics and marketing research exist and are flourishing.

Hey – You're a Forecaster

It was a typical Wednesday afternoon at a previous employer when my boss burst into my office, flung some papers on my desk, and announced, "We need some forecasting numbers to give to manufacturing. Oh, and there's a meeting Friday to discuss this." And, of course, he scurried away before I could ask any questions, as bosses tend to do.

This was not good news. I did not want any more responsibilities. My job title was intentionally vague, permitting my boss to assign me any task no one wanted or could do. I already did pricing. The pricing function there was intricately complicated by design. When I questioned this absurdity, a company executive told me, "We don't want anyone to understand our pricing." Yet, I was expected to administer this cockeyed strategy and ensure it was correctly programmed into the mainframe. I lived in fear that we would get sued, and I would have to explain our convoluted system under oath. I was worried the judge would not buy my testimony and throw me in jail, and my company would fire me, claiming, "Whoa, what a liar that guy is!" Eventually, when I was relieved of my pricing duties (more on that later), it took two-and-a-half people to do just the pricing part of my job.

Besides pricing, I produced whatever marketing research and sales analysis was needed. In addition, because my boss considered me the most competent staff person, I would receive projects that others didn't want or failed to do, including traveling from coast to coast to interview customers for our testimonial video.

What Is This Stuff?

So now I had yet another responsibility—forecasting? I was already vastly overworked. The flung papers contained spreadsheets listing the product families with historical production data for the past several months and then blank spaces for future months. Before Friday's meeting, I eyeballed the data, made a few guesses, did some pencil-whipping, used some calculations, and voila! In about 20 minutes, I had myself a forecast.

I entered that Friday meeting with just those printed spreadsheets and a pen and had hoped this would be a short meeting. I was aghast when the General Manager, my boss, the Customer Service Manager, and several materials and manufacturing people entered the room. Oh, this meeting was much more elevated than I anticipated.

The door closed, and the meeting began. I was shocked by what transpired. The discussion deteriorated into mass chaos. It was one of the worst meetings of my career. We will revisit this meeting later in the book. I had hoped this was a one-time task, yet the monthly forecast meetings continued. I was now suddenly a forecaster and part of the demand planning process. Yet, I lacked training and knowledge. I needed to learn how to forecast but had no time to spare. Not only couldn't I stand my ground, I didn't even know where to stand!

A Treacherous Job

Forecasting is a treacherous job. It is the only job in the corporation where your performance can be precisely measured monthly (assuming monthly forecasts) and where that performance can be deemed unacceptable by The Cheese (the generic name for upper management used throughout this book) for any reason. At any moment, they can inform you that your performance is not up to standard and show you the door. Incredibly, this can even happen if your performance is good! Once the decision has been made, you have no argument or recourse. You just pick up your stuff and leave.

I do not want that to happen to you. This book aims to make you a competent forecaster, manage your boss and The Cheese, and be a valued, respected employee.

Business predictors are also in the hot seat because you are only as good as your last projection. Adapting the concepts in this book will help you improve your job performance, manage expectations, and advance your career.

A Target on Your Back

Scapegoat – a person who is blamed for the wrongdoings, mistakes, or faults of others, especially for reasons of expediency. *See: whipping boy, fall guy, patsy.*

"But the forecast was wrong!"

This complaint can come from people in manufacturing, purchasing, materials, finance, accounting, and even sales. Someone is taking heat from The Cheese for something that has gone wrong or poor performance metrics and is blaming the forecast. You become a soft target. The easiest, quickest, and, unfortunately, best scapegoat available.

This blame transference tactic is used in many circumstances and often sticks because the forecast is seldom spot-on. If The Cheese hears these complaints repeatedly from various sources, the logical conclusion is, "We really have to do something about that lousy forecasting person." Note that it is rarely "What can we do to help the forecasting person do a better job?"

The Forecast is Never Wrong

From the forecaster's perspective:

The forecast is never wrong; it varies in its degree of accuracy.

From *their* perspective:

The forecast was wrong.

As a forecaster, that criticism is nebulous because virtually every forecast is technically wrong. The probability of hitting the number exactly is akin to picking a winning lottery number. So, is the forecast actually wrong? No:

The forecast is never wrong; it varies in its degree of accuracy.

What *they* are really saying is the forecast is "wrong" when it is inaccurate to the point that it causes them issues that they cannot or did not handle sufficiently. Or, as stated previously, perhaps they screwed up, and they need a convenient fall guy.

What is an acceptable range in the forecast regarding accuracy? You want to establish that standard range (Chapter 8). Creating and measuring against that standard is essential (Chapter 13). This will determine whether an accurate or right forecast is made versus an inaccurate or wrong one.

This is essential because the tendency in a toxic corporate environment is to shift blame when errors are made. As previously stated, forecasters are the scapegoats with flashing red targets on their backs.

Your job is to work diligently to produce the most accurate forecast possible, and *their job* is to take that forecast and carry out *their* responsibilities as well as possible. This should be a team effort. When the team succeeds, you succeed together. When the team fails, you fail together. When you can get people to buy into this concept, they will not scream, "But the forecast is wrong!" whenever things go sideways for any reason.

This book will help you create a forecasting process that will enable you to calmly and confidently respond to someone screaming, "The forecast is wrong!" Even if it is The Cheese. Because:

The forecast is never wrong; it varies in its degree of accuracy.

The key is to know how accurate the forecast is and, if it's not accurate, what factors caused the forecast to fail to meet the accuracy standards. The forecast is never wrong, although the assumptions you make to support the forecast are often wrong (Chapter 11). This information allows you to have a productive discussion about the forecast instead of a heated argument over the numbers and stand your ground.

Sometimes, it is Pass/Fail

For business predictors, the results are often more precise. The stock pick tanked instead of rising. The new product research indicated success, but the product failed. Your recommendation for entering a new market did not work out as planned. Of course, measuring the "no-go" recommendations is always tricky because they never happened. Regardless, it is still necessary to document your decision assumptions and keep a scorecard of your successes and failures. Doing this will make you better at your job and may even preserve your employment.

For forecasters, you become the company scapegoat unless you shield yourself from blame. Historically, the scapegoat was sent into the wilderness after the chief priest had symbolically laid the people's sins upon it (Oxford Languages). This book will keep you from being thrown out into the corporate wilderness.

Operating in Difficult Business Environments

Unless you are employed by Unicorns and Rainbows Inc., you work in a challenging business environment. Company cultures vary greatly, though all cultures contain some degree of toxicity. Companies employ imperfect people, sometimes incompetent people. Add in the profit pressures, work stress, office politics, and the overly ambitious fellows, and the soup can get rather caustic.

The pressure can make you a better forecaster and business predictor. Stress can motivate you to raise your game and become more proficient. Sometimes, you must improve your performance to survive, let alone thrive. The strategies, tactics, and tricks in the following chapters were developed in my attempt to swim with the corporate sharks and to help you stand your ground.

Relevant Story #1

The neanderthals love to criticize your forecasts ...

I was a newly hired product manager preparing to launch my first significant new product. I presented the product concept to the executive staff. I was confident it would be quickly approved. The competition was kicking our butt, and this product, basically a superior knock-off, would be a much-needed winner.

After I finished the presentation, The Cheeses asked for a sales forecast before they would approve the project. I had not prepared one since I considered the decision a no-brainer.

So, I put together a sales forecast and got on the schedule for the next staff meeting. I confidently and proudly projected my sales forecast on the screen.

Now, you might think this request was in accordance with good business practice. Regrettably, that's not why they requested a forecast. It was due to a toxic business environment in which the executives sought any opportunity to torch the serfs.

I sensed trouble when the forecast appeared on the screen. There was nothing wrong with my numbers; it was a reasonable forecast. However, it was like I had thrown red meat before a pack of hungry wolves. They thought: *This rookie is naïve and stupid enough to show us an actual forecast! Let's rip him apart!*

All corporate hell broke loose as the various Cheeses took turns trashing my numbers. It was a horrible experience. They didn't talk about the product or the market, how the forecast could be reached, or even what they thought the forecast should be. They just tore me to shreds. Subsequently, over time and through much effort, I changed the corporate culture regarding forecasting. This book will explain how I did it.

The new product was eventually launched because it was a no-brainer. And, it was successful. The sales actually exceeded my forecast in the end.

You Are Never Really Safe

Even if you consider your work environment relatively calm, as a forecaster or business predictor, you can never let your guard down. The bombshell can come at anytime from anywhere. You are only as good as your last forecast or recommendation. The Cheese, and especially New Cheese, can arbitrarily deem your performance substandard and dismiss you with a wave of the Cheesy hand.

The concepts and processes in this book will help you manage your boss, The Cheese, and the political environment and make you a Master Forecaster and Superior Business Predictor. I want you to be able to swim with the sharks, stand your ground, and flourish in your profession.

Forecasting Must Be a Team Effort

Developing a forecasting process that includes input and communication with others, inside and outside the company, dramatically reduces the conflict. If you live in the corporate silo, you will die in that silo. A forecasting Lone Ranger will fail because when the heat comes, you will lack the necessary insulation provided by having a solid forecasting *team*. To be a Master Forecaster, you must get help from all pertinent departments. A strong team can support you when you stand your ground.

Your co-workers must be prompted to feed you essential information when needed. It is highly frustrating when you find out you missed your accuracy goal because a salesperson should have informed you of a plant maintenance shutdown at a critical customer. Likewise, business predictors should form information networks and communicate consistently with colleagues and stakeholders.

When I forecasted the North American commercial vehicle market at my final employer, I emphasized to our clients that they were part of my forecasting team. Yes, the forecast users need to be involved in developing the forecast. I encouraged customers to inform me of essential market developments I might not be aware of. If I delivered a forecast they disagreed with, I wanted them to call me immediately to discuss it. Why? Because the only thing worse than putting out an inaccurate forecast is not knowing you put out an erroneous forecast!

You Are Still Vulnerable – Even If You Are a Good Forecaster

You are still vulnerable if you consistently put out accurate forecasts in isolation. People will become accustomed to the forecast being correct and expect that performance every month. They may even believe that forecasting is easy or that you are not working hard to achieve your results. You are only as good as your last forecast. You might have been invisible to the upper Cheese for years until that one bad forecast that spins everything out of control. Suddenly, you will be highly visible to all and viewed as incompetent. This can be career fatal. The book helps you survive a poor forecast and stand your ground on the next one.

A "lousy" forecast soon after you get a new boss is especially dangerous. I have found that your status is always tenuous when the person who hired you leaves the company. Your new boss has no vested interest in your success and can dismiss you on a whim. And as previously stated, they don't need much motivation to boot forecasters and business predictors.

Relevant Story #2

Getting noticed ...

At a previous employer, I was responsible for executing price changes. Our top selling item by volume was priced in the system as "each", even though they were described as "Box of 100" everywhere else. Mistakenly, I entered the new prices per box instead of per piece, meaning the prices were 100 times higher than they should be. The next day's sales reports showed the company had achieved record sales by a huge amount. Even though my boss identified and quickly fixed the problem, various Cheses stopped by my desk throughout the day to ask me how this happened. Yes, now they all knew who I was, although not for a good reason.

Move The Target

You work in a job where everyone wants to blame you for their mistakes. Your job performance can be deemed "below standard" by anyone at any time, and you may only be noticed when you mess up. These analytical-type jobs can be nerve-wracking because delivering the wrong numbers can have enormous consequences.

As previously mentioned, you have a target on your back. How perilous this condition is, depends on your company's political environment. If it is polite and orderly, like, let's say, Canada, you don't have much to worry about. If it resembles a terrorist state, then you are taking fire constantly.

What can be done about this? You need to move the target from your back to your front. The attacks will still come, except you will see them plainly and quickly. This allows you to block, dodge, duck, blunt, escape, disable, hide, etc. It may even permit you to go on the offensive if needed. It is much easier to stand your ground during a frontal attack versus getting blindsided.

This book aims to make you better at your job and help you flourish in that job. You must move that target to the front of you, and I will show you how that's done.

If you keep the target on your back, a shot you never saw coming will land dead-center. Sadly, you may never even know who pulled the trigger.

Conversely, with the target on your front, you can become a Master Forecaster. You will become a master at corporate Jiu-Jitsu. And the wonderful thing is, as soon as those hacks in your company experience your defensive skills, the attacks on you will diminish. I want to turn you into a black belt Master Forecaster, or Superior Business Predictor. It's time to learn how to stand your ground – cue the Tom Petty music. Take a bow and get ready to rumble.

Chapter 2

The Relentless Pursuit of Accuracy

Lexus commercials touted the company's Relentless Pursuit of Perfection. In the forecasting/analytics game, this is not achievable. Your forecasts, projections, and analysis will never be perfect, so let's replace this with The Relentless Pursuit of Accuracy.

The bad news is that your forecasts/projections will never be perfect. The good news is that it leaves much room for improvement. Quality Assurance people would refer to it as continuous improvement. Of course, continuous improvement in dynamic environments like ours is like chess versus the "checkers" of most manufacturing situations.

To become excellent at your job, you must be committed to the relentless pursuit of accuracy. You must never be satisfied with your current performance. Yes, you can bask in the glory of that last impressive forecast or analysis. But remember, you are only as good as your last forecast. That may have been a great, spot-on forecast, but now it's in the past - time to focus on the next one.

The Continuous Improvement Paradigm

To relentlessly pursue accuracy, we can borrow some principles of the Six Sigma concept of continuous improvement:

Create an accuracy standard. (Define)

Measure your current performance versus the standard. (Measure)

Determine the gaps in your forecasting/analysis process. (Analyze)

Make changes in your processes to improve your accuracy. (Design)

Measure your accuracy. (Verify)

You are not manufacturing a product, but you are creating a forecast. The process presented in this book will show you how to use the steps above to continuously improve the accuracy of your forecasts and analysis/ recommendations. To become a Master Forecaster or Superior Business Predictor, you must:

Measure – Make Design/Process Changes – Measure

to achieve continuous improvement.

Because you are one of the few people in the company whose performance can be precisely measured every month, it is essential that you are committed to continuous improvement and are measuring your accuracy.

Personal Performance Standards

Chapter 8 explains how to create forecast accuracy standards for product families. These standards will be public and visible to your stakeholders inside the company. In addition to the public standards, it is beneficial to create personal performance standards.

This personal standard will motivate you to improve your performance. I would always measure my performance (covered in Chapter 13) as soon as the previous month's data was available. That day, I would feel great, satisfied, or disappointed, depending on how accurate my forecast was. But the critical result is that I would feel something. To be a Master Forecaster, you must care about your job and consistently produce accurate forecasts.

If you have a good year, you may choose to share your personal performance results with your boss in your performance review. If you have a bad year, keep those results quiet unless asked. This is not disingenuous. You have a difficult job, and this is a game of survival. However, the forecasting process discussed later will provide ample information to explain why your performance may have failed to meet expectations when needed.

Performance reviews are always tricky. Often, your boss will attempt to downgrade your performance to limit your salary increase or for other irreverent reasons. Forecasters and Business Predicters are especially vulnerable here, especially if they start the discussion with, "Your forecasts are always wrong." You can then remind them of the forecasting standards you use and how you measure against them. This may improve the final write-up, but sadly, it probably won't be good enough to get you more money.

If you keep a personal scorecard, you may not want it accessible to everyone in the company, so keep it off the network. Unfortunately, some people spend work time scouring the personal computer files of others. Due to those computer snoopers, I once heard through the rumor mill that one coworker padded his resume by listing "Rocket Scientist" as a prior position, while another claimed to have a very fake Ph.D.

Business predictors have more difficulty setting hard performance standards but should also be committed to continuous improvement. You can become more proficient at your job by following those Six Sigma steps previously listed, whether you are recommending stocks or conducting marketing research studies. In the Business Predictor realm, your big wins and big losses will be obvious. It is best to keep a scorecard of all analyses, studies, and recommendations.

Knowing your score can still be highly beneficial at specific points. Like when you make a bad call:

"Wilson, you really missed on that one!"

"Yes, I had been on a hot streak, but that one didn't go the way I anticipated."

Documenting your results will also help during performance appraisals when you have had a good year, but your most recent call was faulty. Of course, if you have had a bad year, you never show them those stats and hope for the best.

Relevant Story #1

When performance reviews go badly ...

You need every weapon at your disposal during performance reviews because many go horribly.

I had planned to take the week off after my second daughter was born. But she was three weeks early, and the product launch I was leading was two weeks late, and Shazam, they ended up on the same week. With my wife's blessing (my mother-in-law helped out at home), I made the difficult decision to work that week and manage the product launch.

Just two months later, during my yearly performance appraisal, I received a mediocre review. But what burned the most was the "needs improvement" on the "Committed to the Job" metric. That was the most bogus rating I ever received on a performance review.

Continuous Improvement Requires Continuous Learning

Micro-Learning

The relentless pursuit of accuracy requires continuously analyzing the market environment. You need to be learning something new every day. This is not tedious work. It's something I enjoyed as a forecaster. The relentless pursuit of accuracy means you should never be bored in your job.

Chapters 9 and 10 detail how to build your information network. In the pursuit of forecasting excellence, you read articles, talk to people inside and outside the company, and analyze new data constantly as part of your job. At a previous employer, I analyzed the daily order report for trends and indications of how my forecast was looking for the month. I kept my finger on the pulse of the market to detect demand trends at the earliest stages. If The Cheese stopped by and asked how we were doing, I had an immediate answer.

How much time you spend tracking your market depends on its dynamic structure. Some industries/sectors move so fast that daily knowledge is critical, but daily learning helps you achieve forecasting excellence even in stable environments.

Relevant Story #2

Most markets are more dynamic than they seem ...

People from outside my industry thought it wasn't difficult to forecast the demand for commercial equipment – my final job. I would explain it this way:

Let's say there are 50 economic/industry factors (I never tried to count them) that can potentially impact equipment demand. At any time, five of those factors are the key factors. So, you study those factors intently until you think you know what is happening and forecast based on those factors. However, over time, the business environment changes. When doing a subsequent forecast, you now find that two of the factors remain important, two others are now unimportant, and the other one is losing relevance. Therefore, you must identify the new factors driving demand and reevaluate the previous group – a very challenging task, indeed!

What looks like a stable market that is simple to forecast is actually highly dynamic and a forecasting challenge. In fact, the commercial equipment markets have some of the broadest demand swings in the industrial economy.

The Most Informed Person in The Room

You may not be the smartest person in the room, but you should be the most informed about the market environment and demand drivers. You should know as much about this as the CEO.

Why? The culmination of what is happening in the world, country, economy, industry, and company is compressed down to two simple numbers: sales and production. Simply, these are the very two things you are tasked with forecasting. To forecast it all, you have to know it all! Okay, no one knows it all. But if someone knows more than you, they have an advantage over you, whether it be someone in your company or the competition. Information is power, and having much information helps you to stand your ground.

And it is essential to be well-informed to handle this type of situation:

(In a meeting)

Hotshot from Finance: "Your forecast is garbage because I heard that exports are supposed to fall due to the stronger dollar."

You: "Yes, I saw that economic report, but we are still in decent shape because we have some price protection in the countries we sell to the most, and the forecast reflects that."

And be very cognizant of what questions are asked by The Cheese. It reveals which indicators and factors are important to them. Strive to become an expert in those areas. That way, when asked again, you can respond with an astute, credible answer and look like an expert!

Macro-Learning

The relentless pursuit of excellence also involves outside personal development to boost your forecasting knowledge and skills. These include:

Book and Journals – Hundreds of books on the statistical nuts-and-bolts of forecasting exist. The challenge is finding those most relevant to your industry and forecasting needs, but keeping your skills sharp and up to date is worth the effort. *The Journal of Business Forecasting*, published by the Institute of Business Forecasting and Planning, is the best source of current forecasting trends and techniques.

A wide variety of books also exists for business analysts/predictors. They range from basic *Business Analysis for Dummies* to those featuring high-level statistical analysis. There are several journals for business analysts and an International Institute of Business Analysts.

Seminars, Webcasts, and Conferences—I recommend trying at least one of the seminars, conferences, or webcasts offered in your discipline and then ascertaining their value. In forecasting, the Institute of Business Forecasting and Planning holds conferences, and several firms provide practical training and events.

And if your industry provides conferences on demand drivers/trends pertinent to forecasting or prediction, Go! You will gain insight from different perspectives and make important industry contacts, which can help you become an expert in your field. Being well-educated and skillful in your craft will also help you stand your ground.

The Benefits of the Relentless Pursuit of Accuracy

Better Forecasting/Predicting

In the long run, the commitment to improving accuracy should improve your job performance. It allows you to become proficient in your field, maybe even being considered an expert. This level of performance can help you attain larger raises and promotions and provide greater job security.

If your company doesn't appreciate your commitment to accuracy or dumps you out on the street after ten years of outstanding performance (the very end of the book explains this one), the skills you have developed are marketable and transferrable to other industries.

Becoming a proficient forecaster instills a sense of professional pride. Forecasting is a difficult task - more difficult than others realize. Pursuing excellence and knowing you have achieved proficiency boosts your personal ego and provides job satisfaction. This allows you to promote yourself and display your forecasting accuracy results with confidence and stand your ground.

Do Not Sit Still

Some business gurus will tell you if you are not growing professionally, you are dying. This is true in the forecasting/prediction realm. I would ask this: "What are you doing right now to improve your accuracy?" If the answer is "nothing," then you are vulnerable to the arrows and attacks warned about in the previous chapter. Fortunately, this book will provide you with methods to improve your accuracy.

This is a race, a race towards excellence. The relentless pursuit of accuracy. You must be a runner, not a spectator. And the pace is constantly changing in the era of mega-data and exploding technology advances. You must run the race faster than ever to keep up.

The relentless pursuit of accuracy demands that you learn something new every day, have the most knowledge in the room, and are committed to continuous learning and personal development.

Do not be satisfied with your current level of forecasting/prediction accuracy. You cannot sit still and stand your ground. Be committed to continuous improvement in both your business and personal development. When you do, you will rise to the top of your field and have the opportunity to be an expert, even a superstar in your industry. And don't hesitate to promote your achievements along the way.

Chapter 3

The Heart of a Forecaster

Forecasters must approach their jobs differently than others to flourish and survive. Here are some intangibles I learned and developed over my career. They espouse the attitudes and tactics beneficial to being a Master Forecaster/Superior Business Predictor.

Some business gurus will no doubt disagree with some of my concepts. That's fine; these are my personal rules that I pass on for your benefit. And besides, these experts have never had to face a flaming Cheese demanding that you explain how you missed the forecast.

Predicting the Future

"Prediction is very difficult, especially about the future."- Niels Bohr

Your job is not that different from a fortune teller – you both get paid to predict the future. The fortune teller uses all the tricks of the trade to deliver a believable story. You use all the data, information, and experience to develop a credible forecast.

Forecasters don't get the respect they deserve. While it is acknowledged that forecasting is challenging, people expect an accurate forecast every time, and as previously stated, you can go unnoticed until you screw something up.

Although forecasting is a challenging profession, most forecasters are underpaid for the value they provide. However, business predictors, especially stock pickers, have the opportunity to rack up huge earnings. Sharp market researchers at large corporations also have high value and are compensated well.

That's your job - somehow accurately predict the future every time and get roasted when you fail. The fortune teller has an easier life. But that's your mission, so let's just focus on becoming that Master Forecaster who can stand your ground.

Taking Ownership of the Forecast

Many forecasters do not *take ownership* of their forecasts. They attempt to separate themselves from the numbers to avoid being criticized and punished.

Because forecasting is so difficult and to avoid becoming the *scapegoat*, it is tempting to develop a forecast and then toss it in the air for whoever wants it, attempting to eliminate any connection to it or responsibility for it. If the forecast is questioned, the likely responses include:

"I really don't know where the market's going."

"I didn't have much time to put into it."

Because few people seek to become business forecasters, many of us end up in the field by chance, being assigned the responsibility somewhere along the way. And like other unwanted tasks, you may not do your best work, hoping they might transfer the responsibility.

When I was assigned the forecasting duties in the marketing department at a previous company, I wanted nothing to do with it. It was given to marketing because that department held the least political power in the organization, and no other department wished to have the "turd". (More on this in the next chapter.)

Now, I had a good reason to dread this new task. I was already overloaded with other responsibilities. I had no interest in doing forecasts and no time to do it. Those early forecast meetings were horrible. Unfortunately, I was stuck doing the forecasts because no one else in the department had analytical skills.

However, this attitude is a recipe for inaccurate forecasts and dysfunctional processes. To become a Master Forecaster, you must own the forecast. Oh, excuse me, I made a mistake there - You must OWN your forecast. YOU MUST OWN YOUR FORECAST.

It's the same principle of ownership present in all areas of life. If you own your home, you treat the property much differently than if you rent it. You care for your own house more than a leased/rented one. When you take ownership of your forecast/prediction, your whole attitude and approach will change.

"There's a leak in the roof. If it gets really bad, we'll call the landlord" vs. *"The roof of my house is leaking; I'll get it repaired now to prevent further damage."*

"Yes, my forecast was off. I'm not really sure why." vs. *"My forecast was out of standard this month. I must research why and tighten my process."*

Again, one reason people don't take ownership of the forecast is to avoid being held responsible for poor forecasts and being blamed for the subsequent results. This strategy may work in the short term, but the result will continue to be poor forecasts and eventually be disastrous.

Unfortunately, when you take ownership of the forecast, you get that target on your back. If the target remains on your back, you are a sitting duck. The goal is to move that target from your back to your front, where you can better fend off the arrows.

The processes, guidelines, tricks, advice, and strategies detailed in this book will enable you to fully own your forecast and defend it from attacks on all sides. This will result in consistently accurate forecasts, greater job satisfaction, and career advancement. If you don't own the forecast, you won't be able to stand your ground.

There is an "I" in this Team

Forecasting should be a team effort, and building that structure will be covered in subsequent chapters. However, no matter how many people are involved in the process, somebody still needs to own the forecast. No matter how it was developed, there is still one owner - you.

Relevant Story #1

In the end, the forecast belongs to you.

At my final employer, we utilized a team forecasting approach. The forecast meeting consisted of six people analyzing, discussing, and debating the numbers. Different scenarios were considered, and various opinions were rendered.

Ultimately, the team finalized the forecasts. However, as soon as they were published, they became my forecasts. I totally owned them. I was responsible for supporting, communicating, advocating, and defending them.

Of course, the communication to customers and the industry was that it was the Company Forecast because it was detrimental to imply that only one person worked on it, which wasn't the case anyway. But functionally, it was *my* forecast.

This ownership mindset also sometimes affected my discussions during the forecast meeting. When, on rare occasions, I would disagree with the group consensus, I would interject, "Remember, I'm going to have to explain and defend these numbers at the next webinar." Ultimately, whether I fully agreed with the forecasts or not, it was still my job to own them, present them, and defend them.

"I Don't Know" is Not an Acceptable Answer

You will be asked more questions as you gain more knowledge about the market/industry/company (covered later). The more you know, the more you will be asked, and this is a good thing, as people depend on you as a valuable source of market knowledge. The Cheese and others will start asking you questions during important meetings. You should be able to handle most questions, but some will stump you.

These questions fall into three categories:

1. Great questions on issues, factors, or analyses you had not considered before.
2. Questions that cannot possibly be completely answered.
3. Questions that, well, I won't call them stupid but have no relevance or value to the subject.

You could quickly answer "I don't know" to these questions and hope the discussion moves to a new subject. But I believe that "I don't know" is not an acceptable answer. I know that business gurus, social scientists, and even ethicists might disagree. Still, there is a better way to handle these questions, which can build your credibility and strengthen your political ties in the corporation.

To answer the questions listed above:

For the first type (Legitimate questions):

"That is a great question. I have not looked at that before. Let me dig into that and get back to you."

Or, if it is more basic – *"Good question; I'll look that up and get back to you."*

For the second type (Nearly impossible questions):

"That is a tough question. I'm not sure if that data is available. I can see if there is any related information out there."

For the third type (Irrelevant questions):

In this situation, use the politician's ploy of answering a different question than what was asked. A question you can answer. Just be sure that your answer is related in some way to what was asked.

If the questioner doesn't buy your answer and repeats the question, shrug your shoulders – note that you are not saying – "I don't know." You are indicating that was your best answer to the question. However, if the person is The Big Cheese, your final answer is always, "I'll look into it, sir."
This strategy permits you to present yourself as knowledgeable and competent. You are the person who *knows stuff* versus the guy who says "I don't know" a lot. Getting back to the person and providing relevant information builds your credibility.

Therefore, strike the term "I don't know" from your vocabulary.

Any Answer Is Often Better Than No Answer

This one is trickier. My philosophy, which many people, including some former coworkers, disagree with, is to give the best answer you can at the time, even if there is a chance it is incorrect. Under challenging circumstances, such as presentations, webinars, and media interviews, the best choice is to wing it. Give the best answer you can come up with, delivered with confidence. And be sure to add enough qualifiers to leave you with an opening large enough to drive a semi-truck through.

Think of it this way. If you are the expert and don't know the answer, chances are that 99% of the audience doesn't know the correct answer either. And the one person who knows the correct answer may even reconsider their viewpoint after hearing your response.

If your answer is incorrect:

1. Virtually no one is going to know.
2. The answer will probably not be challenged at that moment.
3. If your statement turns out to be wrong in the future, few people will remember what you said.

Of course, later, you will quickly research the correct answer, so if you are incorrect, you are prepared to provide the right answer if you are challenged or if the question is ever repeated.

If your answer was faulty, admit you were wrong if questioned. "Yeah, I did review the subject, and it turns out that …."

Relevant Story #2

Winning the political games.

When I began building a forecasting process at a previous company, The Big Cheese noticed my progress, and I was asked to give short market updates at executive staff meetings. At the end of my update, the various Cheeses would ask questions. At first, this seemed totally legitimate, but I soon realized they were actually playing a game that I call "Stump the Lackey".

The first couple of questions were usually basic; then, someone would ask a tricky question, to which I would reply, "I don't know." In "Stump the Lackey", if you can get the Lackey to say, "I don't know," you are a winner and have proved your knowledge superiority! Some weeks, this game would have two or three winners, as they took turns stumping me.

I soon became frustrated with this game. So, after the meetings, I would research every question I couldn't answer. Some questions were irrelevant, but I learned a lot through the research. However, I played my own game. I didn't get back to the people who asked the questions when I found the answers. I waited until the question was repeated in future meetings, and then I delivered the goods.

Eventually, due to my knowledge-building process (Chapters 9 and 10), I could answer almost every question thrown at me during the meetings. I continued to research anything that stumped me.

Now, in the game of Stump the Lackey, if you ask a tricky question, but the Lackey answers it, you lose – and the game is no longer fun. So, the number of questions steadily declined until only a few genuinely relevant questions were asked - and I knew most of those answers! I was no longer a lackey but a lean, mean, knowledgeable forecasting machine!

I ultimately benefitted from playing this game. It motivated me to research and gain valuable knowledge. But it was still one of those silly political games played by immature, puffed-up jerks. The light came on for me when the questions dried up. If they really cared about my answers, they would have continued to ask questions because I had become more knowledgeable. When they could no longer stump me – literally, game over.

Saying "I Don't Know" Can Have Bad Consequences

When the Big Cheese at one of my previous employers asked you a question, he thought you *should* know the answer. It was a test. Unfortunately, if you flunked this test enough, you failed to keep your job. One poor sap got booted soon after uttering a casual "I don't know" in an important meeting.

So, my standard reply to the Big Cheese when I didn't know the answer was, "Let me get back to you on that." And I always did. This had a profound effect on my approach to complex questions. Years later, at another previous employer, I literally flinched in my chair when the warehouse manager delivered a dismissive "I dunno" to a question on a metric he should have known off the top of his head. The corporate culture there was obviously very different.

Therefore, when you are in a public, real-time situation, you give the best answer you can. When you are in a private situation or have more time, you research the answer. Knowing the answers helps you stand your ground.

Bottom Line: Master Forecasters don't say "I don't know" – they learn what the correct answer is.

Never Say the "G" Word

Because your job is similar to a fortune teller, your coworkers may assume you lack credibility, and you can get the following comments regarding your arduously prepared, well-researched forecast:

- "So, you just had to guess this month, huh?"
- "So, that's your best guess?"
- "Nobody knows, so it has to be a guess."
- "What's your best guess on this one?"
- "It looks like you guessed wrong last month!"

To each of these statements, your immediate reply should be:

"I never guess; I deduce based on the best information I have."

Because: **YOU NEVER GUESS – YOU DEDUCE BASED ON THE BEST INFORMATION YOU HAVE.**

Deduction (def) – The deriving of a conclusion by reasoning or logic. (Merriam-Webster)

Or:

To reach an answer or a decision by thinking carefully about the known facts. (Cambridge Dictionary)

Deduction sounds much more impressive than guessing – don't you think?

Therefore: **YOU NEVER GUESS – YOU DEDUCE BASED ON THE BEST INFORMATION YOU HAVE.**

And you don't want to guess or ever imply that you do. Because anyone in the company can guess. Do you want management to believe anyone can do your job? Carl, the accounting clerk sitting in the corner with no social skills? Yes, Carl, the guy who dresses sloppily and picks his nose, can also pick numbers out of the air. Do you want Carl to get your job?

Every time anyone implies that you are guessing, even The Cheese, the reply has to be the same:

"I never guess; I deduce based on the best information I have."

Some of the screwballs may try to argue the point, but you will have the facts (covered later) to back up your position.

Your forecast may indeed be based on highly variable outcomes and thin data. However, you will state what assumptions and data your forecast is based on and how that might impact the accuracy. People may conclude that the basis for the forecast is so wobbly that you have to guess. But that is not your position:

"I never guess; I deduce based on the best information I have."

This is also the same position for business predictors. You are not going to guess at your conclusions and recommendations. You may be dealing with thin or unreliable data and highly variable circumstances. You may have to make some shaky assumptions. But you are not guessing, so the statement above still applies.

It's the "G-Word"

Don't even say "guess" when discussing your work. Treat it as the "**G-word**" for forecasters and prognosticators. You never want to guess or imply that you do, so never put the word in the air or on the table. You can't stand your ground if you guess.

One of my colleagues at my final employer had no problem freely using the "**G-word**" during his webinar presentations. He said it to indicate his data was inconclusive, making the forecast highly variable. He would say the data was so murky that it was his "best guess." I would protest vehemently afterward. My main argument was this:

"Our customers pay us to forecast. They do not pay us to guess. They can guess on their own. If we are guessing, why do they need us?"

Relevant Story #3

Never guess – no matter how shaky the data is.

And, if you can't forecast accurately, forecast often.

In March 2020, COVID-19 shut down the U.S. economy. This put enormous pressure on my final employer to deliver our April industry forecasts to our customers by the end of the month. The virus's expected impact on the economy changed daily. In response, we met five times in ten days before agreeing on our economic forecast in an unprecedented Saturday morning meeting.

My responsibility then was to finalize the commercial equipment forecasts that were due by the end of the day on Monday. The process was to run the economic forecast to get a freight forecast and then run the freight forecast through a highly complex model to get a baseline equipment forecast. Then the team met to analyze industry data and qualitative information to formulate our assumptions and adjust the numbers off the base forecast.

However, we had no reliable data or information to make any assumptions. Any adjustment at that moment would be a pure guess. But:

"I never guess; I deduce based on the best information I have."

Therefore, I decided that the forecast would be the exact number generated by the model, with no adjustments, and we would explain this to our customers. We delivered the forecasts on time, and our clients respected our transparency. Ultimately, my final employer always boasted our forecasts are data-driven. We had remained faithful to this mission under extreme pressure.

That main forecast turned out to be 22% too low. That's not bad, considering the less dire circumstances later (we didn't all die) could not have been predicted in March 2020.

Personal Biases Can Pollute the Forecast

Be careful to keep personal biases out of your forecasts and analyses. Any factors outside the information, data, and the process, pollute the forecast. The projections and prognostications are the results of your analysis, not what you hope will happen.

Politics has invaded all aspects of business and society. Still, it has no place in your forecast. Over the past decade, I have seen respected forecasters fail in their mission due to political bias. The news media has even tagged some economists as either Democrat or Republican. You sacrifice accuracy and credibility if you get too political. However, you can get on television much more often.

If you are marketing or presenting your forecast/analysis outside of your company, be as objective as possible and be aware of your customers' or audience's political bent. Leaning one way when your customers are leaning the other can cause significant problems.

Likewise, stock analysts should not make calls based on personally liking the company's products, management, or political preference. Market researchers' analysis should be free of personal opinions of the product, topic, or subject.

Your forecasts and analyses are based on the best data and information available – not on hope or want. Biased forecasts and predictions are often inaccurate.

Relevant Story #4

Objectivity is so much better than bias.

During the months before a presidential election, a respected colleague wrote a series of monthly economic commentaries that assumed their favored candidate would win in a landslide and this would greatly benefit the industry and the economy.

I was responsible for the final edit of these articles. I made the difficult decision to delete all the biased comments. On one commentary, I stripped out almost half the content. I was uncomfortable doing this much cutting, but I wanted to keep our analyses as objective as possible, and our customer base politically leaned in the other direction.

However, my decision was validated after the election. Not only wasn't it a landslide, but my colleague's favored candidate lost. We would have looked stupid and partisan had I not made the edits. This was one of the best decisions of my career.

Admitting When You Make the Wrong Call

While the *forecast* is never wrong, your assumptions will often be faulty, based on all the variables and unknowns in your market analyses. Business predictors are often wrong on their go/no-go calls. People should not expect you to be perfect, but they should expect you to be competent.

If you are defensive when your predictions are inaccurate, you lose credibility. People will either perceive that you don't know you are erroneous or won't acknowledge it. And if you don't acknowledge it, chances are you won't research it. If you don't analyze your miss, you will likely make the same mistake in the future.

Therefore, the appropriate response is:

"Yes, I missed that one."

And then either:

"My assumption about (whatever it was) was off, and this impacted (whatever)."

Or:

"I'm still looking into the reasons."

People may continue to challenge you. If they express their opinion about why you were off, it is a productive conversation, and they are essentially partnering with you to find an answer to the question. This is highly beneficial, even if you disagree with their outlook.

However, some people will insist and even delight in pointing out your error. They may even do it publicly in front of The Cheese. Of course, they may even be The Cheese. These people are either your enemies, rivals, or just plain jerks. Every company has a few; I know you are thinking about who they are in your organization right now. The best move is to repeat the statements listed above. You can even ask them, "Why do you think the forecast was too high?" This usually quiets the screwballs right away.

I do have a personal rule about repeating myself for a third time. If someone disregards or ignores your statements twice and insists on babbling on, they either are not listening to you or enjoy beating you up. Either way, it's time to end the conversation the best way you can. Sometimes, I would explicitly state that I would not repeat my statement again.

Yes, admitting your error when you miss a call or forecast is the best strategy, but you seldom need to apologize (explained in Chapter 18). Chapter 15 covers how to fully defend your forecast.

Relevant Story #5

Please listen before becoming defensive.

I found an error in a report from our data/industry analysis provider and called them about it.

"There's an error with the forecast number on page 12," I began. Before I could continue, the analyst gave a long-winded explanation of why the forecast number was correct and why I should accept it as stated. His response was much too defensive. If you are confident of your forecast number, you do not have to be defensive or do a hard sell.

He finally stopped and waited for my response. "Uh, no. You have a math error," I explained. "If you add the first number in the column with the second number, it doesn't equal the third."

There was a long pause as he literally did the math.

"Oh, yes. Thanks for pointing that out," he stammered.

This company lost much credibility with me due to that one conversation. I never had a great deal of confidence in them going forward.

Relevant Story #6

You only get to blast me twice ...

As I walked by, the Finance Cheese called out from his office. As I stood at the doorway, he began to berate me about something related to the forecast. I explained what had happened and why. He then ripped into me a second time without expanding or advancing the conversation in any way. I then repeated what had happened and why it had occurred. I added some ideas about preventing the issue in the future. He then began round three of the torching.

In this case, telling him I wasn't repeating myself a third time would have been detrimental. This would have led to a public shouting match with a Cheese. This guy was a complete jerk, and I wouldn't have been able to out-shout him anyway. But at this point, it was just him continuing to yell at me. However, this wasn't a planned meeting – I had not been summoned. And I wasn't sitting in his office. So, while the hothead was still yelping, I walked away.

One coworker who had witnessed the encounter questioned my actions. I explained that you could criticize my work, but I wouldn't continue to be a punching bag. There were no repercussions. Of course, it helped that everyone thought this guy was a hothead, and my boss disliked him.

Internal Customers

The term *internal customers* has been overused; however, it remains highly relevant in the case of forecasters and business predictors. To do your job well, you need to know:

- The data/information needed.
- In what form the data/information is needed.
- When/how often is it needed.
- What are the accuracy expectations.

It would be best if you strived to keep your customers happy, which, as in all businesses, you will fail to do so at some point. However, satisfied internal customers will help you deflect criticism from other employees, including The Cheese. You know your internal customers are happy when they jump in to defend you before being asked. The concepts and strategies in this book are designed to increase the satisfaction of your internal customers. Happy internal customers help you stand your ground.

I Can't Do That vs. What Can I Do

Often, you will be asked to develop an analysis or forecast that is extremely difficult or even impossible to do. The request may come from an internal or external customer, or more likely from The Cheese, who are prone to ask you to find out anything.

Your natural reaction is to blurt out:

"I can't do that – there's no way to get to that number, analysis, forecast, study, etc."

You react this way because:

- It is truly impossible to do.
- It might be possible, but the work or cost involved is not worth the benefit received.
- You may be able to get to the number, but it has a low value.
- You could get to the number but are severely overloaded with work. Here, you are making the cost-benefit analysis based on your workload.

Rather than outright rejecting the request, it can be beneficial to ask yourself:

What can I do?

Okay, so I can't deliver what the person wants, but what can I provide that is relevant to their request? You should ask some follow-up questions to determine what is behind the request. Flipping the question around tends to stimulate the creative juices. You can often provide valuable data even though you could not meet the original request. This can score some political points with your boss and The Cheese. Considering you have that target on your back, you must build up all the goodwill you can get. And it sounds much nicer than snapping back, "I can't do that."

Relevant Story #7

Not as difficult as it seems ....

One day, my boss told me about a customer's ridiculous request for information. He had not involved me because the request involved something outside my domain. He explained the problem and said, "I told them it was impossible to get to that number!"

Of course, I enjoy a challenge. I went back to my office and pulled some data off the Internet. I combined it with some of our internal data. I then asked the product engineer some relevant questions to tighten the range. After less than an hour of work, I delivered the numbers to my boss, with a short explanation of the process and assumptions, and he gave it to the happy customer. (How to make difficult estimates is covered in Chapter 20).

The Impossibility Diversion

People often tell you something is impossible to estimate, predict, forecast, etc. Frequently, that is correct, but not always. I was a product manager, and we launched a line of private-label tool cabinets. The cabinets contained drawers of various heights and could be custom-designed. In discussing the cabinets, everyone inevitably asked: "How do you determine how many drawers of what heights can fit in a cabinet?" The manufacturer's rep told us that it was based on a complicated mathematical formula, and you were wasting your time trying to figure that one out because it was so complex.

One afternoon, before leaving on vacation, I had some free time. So, I got the product specifications and a calculator. I'm not a mathematician, but it took me 15 minutes to crack the code. I'm unsure why they didn't want to reveal this, but their answer had to be a diversion.

At various times at my final employer, we were asked to forecast or estimate things in areas where data was severely lacking. When my colleagues concluded that we would be unable to provide a forecast, one of my signature sayings was, "I can forecast anything!" Of course, I didn't guarantee the level of accuracy. But sometimes, it is worth the effort to try.

You will sometimes be asked to do the seemingly impossible in forecasting and business predictor positions. One time, when I was having difficulty coming up with a number requested by my boss, I came up with this:

"When asked to do the impossible – sometimes I fail."

Chapter 4
Develop A Process – Trust The Process

"Patience and trust in the process will lead to extraordinary results." – Grant Cardone.

If you are doing forecasting/business predictor work, you already have a process to get to your numbers/analyses. It may be loose and flexible, and you may often cut corners to save time. However, to improve forecast/prediction accuracy and defend it, you need a tight process that eliminates as many *leaks* as possible. Consider me your personal forecasting trainer; I will help you tighten and improve your flabby process.

As mentioned, the challenge is continuously improving your process until you are a Master Forecaster/Superior Business Predictor. Think about your current process or the lack of one. What problems result from your methods? This book will provide the tools to build a solid, effective process that can be modified based on your specific needs. Once you create the process, you can continue to tighten it and watch your forecast accuracy/correct predictions improve along the way. That's when the magic happens! An effective process enables you to stand your ground.

Ground Zero – When There Is No Process

Now, let's go back to my first forecasting meeting. My boss had tossed the spreadsheet to me with minimal guidance, and I pencil-whipped some forecast numbers, totally unprepared for the meeting.

I expected this meeting to be short and sweet. It would be held in the manufacturing conference room, which was small, dingy, with worn-out, uncomfortable chairs. The important meetings were in the new, spacious, cushy conference room near the executive offices.

I got to the meeting and was surprised to see my boss there. Also attending were the customer service manager and three people from manufacturing/materials. We didn't start on time, and I could tell we were waiting for someone. Suddenly, the Big Cheese hurried in and shut the door.

It was my first forecasting meeting and one of the worst meetings of my entire career. The materials manager began complaining about late order changes and a lack of warning about large orders. Then the customer service manager fired back about the recent late deliveries and upset customers. The materials guy jabbed back, raising his volume about the poor communication. The Big Cheese had heard enough. He began yelling at everyone, which was especially frightening since we were all tightly packed in the small room.

It was a total $h!+ show. Then, there was more blaming and finger-pointing until the Big Cheese unloaded a second outburst. The bickering continued until The Big Cheese delivered a third tirade and rushed off. During the tsunami, I held that paper with my forecast numbers tight in my hand, hoping I would not be required to speak because if I did, I was sure I would garner everyone's wrath. But there was so much vitriol that we never even discussed my numbers. I quietly passed my forecast to the materials manager as people hurried out of the room.

Unfortunately, the above example is not unique. Forecasts can generate much unproductive anger in companies, but it doesn't have to be that way. But without an effective process, that target on your back becomes ground zero for the fury. It's easy for a forecaster to become the punching bag in the corporation. When everyone gets angry and frustrated with supply chain/shipping issues, they often unleash their outrage by taking a shot at the easiest target - you.

Thus – A Process Begins

After that disastrous meeting, I hoped it would be a one-hit wonder like many other business trends that flew in with much fanfare but quickly dissolved. But I was invited to the next forecast meeting in the same dingy location the following month. Soon after getting the notice, the materials manager stopped by and flung the new spreadsheet on my desk. However, there was no feedback whatsoever on my previous forecast numbers.

I spent about the same amount of time on this forecast as I did the previous month. We never even discussed my numbers in the meeting, and I received no subsequent feedback. So, I again reviewed the monthly sales data and pencil-whipped some numbers.

The second meeting resembled the first—lots of bickering, blaming, and finger-pointing followed by some loud lambasting by the Big Cheese. Again, my forecast numbers were not discussed, and I considered that a good thing.

The awful monthly meetings continued. I never slept well the night before a forecast meeting, and my forecast process stayed the same. I had little time to devote to forecasting and had received no feedback or direction on my numbers. However, I could not escape being noticed in the meeting forever. At some point, we actually discussed the forecast. What a concept! Unfortunately, this exposed me to criticism from manufacturing and the Big Cheese.

I had protection because my boss was there to deflect and neutralize any arrow shot my way. My boss's sole interest in attending the meeting was to guard me and ensure I wasn't given even more forecasting tasks (see Relevant Story #1 below).

Therefore, we now had a forecasting process. It was ineffective, dysfunctional, and inaccurate, but it was a process nonetheless.

Relevant Story #1

But I don't want to forecast ...

While writing this book, I had coffee with my old boss at that company, the guy who threw that spreadsheet on my desk and attended those terrible forecast meetings. He explained that he never wanted responsibility for forecasting because he did not consider it a marketing function. There had been no discussion with him about taking on this task, so he tried to ignore it as much as he could. He hoped that if we did it poorly, it might go away—a common, yet unproductive, business strategy.

That reminded me of something he had said in a department meeting long ago. He would often repeat illustrations he had heard but get them all jumbled up. We would have to keep from laughing while he seriously explained the concept. Incredibly, when he finished, it somehow strangely made sense. So, one time, he said this:

"Everyone has a turd in their pocket. But they don't want the turd in their pocket because it stinks and it's a turd. They walk around with this turd in their pocket and try to give it away to someone else. But no one wants to take the turd because they already have a turd in their pocket, and they don't want another one. So, you have to try to give the turd away to someone who doesn't already have a turd. But you still have to get that turd into their pocket without them realizing you put the turd in their pocket."

My old boss considered the forecasting function *the turd*, and firmly believed it did not belong under the marketing department's domain. He understood that doing forecasts in that toxic culture was a no-win, in that he would not get credit for accurate forecasts but would always be open to intense criticism when things went wrong.

Beyond that, I was the only person in the department capable of forecasting and was already much overbooked. He didn't want me to *waste* time doing forecasts, so I continued to pencil-whip it every month, and he didn't care. I had no idea if my forecasts were accurate or even beneficial. We repeated this dysfunctional, useless process month after month.

Where Does Forecasting Belong in The Organizational Structure?

My boss thought Marketing should not do the forecasting, but the Big Cheese decided it should. This raises an important question: Where should the forecasting function be located within a company?

The Institute of Business Forecasting – using the IBF Benchmark (a survey from 2016) provides this breakdown. Where Forecasting Function Resides:

26% - Operations/Production

16% - Sales

14% - Separate Forecasting Department

11% - Marketing

10% - Other

9% - Logistics

7% - Finance

7% - Strategic Planning

Every one of these options introduces functional bias into the forecasts. Even if a company is large enough to have a separate forecasting department, there will be some bias based on the Forecasting Cheese's strategy.

Functional Biases

Manufacturing may have a high or low bias, depending on whether they are evaluated on materials inventory or on-time production. The person doing the forecasting will be subjected to pressure from plant managers and production planners.

Sales will either low-ball the forecast if it is used for sales quotas or inflate it to increase sales through higher customer satisfaction and faster lead times. A forecaster based in the Customer Service department will always forecast higher to reduce customer complaints about late deliveries.

Accounting/Finance is subject to various pressures on the forecast. They often decide the desired final number and then work backward to get the forecast to match it. They are susceptible to the most pressure to keep the forecast low by the Big Cheese, especially if the company is publicly held. (More on this later).

Marketing has some biases, especially for new products or if significant cash has been plowed into advertising campaigns. There can also be some bias pressure from Sales, especially if The Cheese is over both functions.

The Best Place?

The three factors in play here are bias, conflict, and accuracy. A separate forecasting department would have the least amount of bias. However, it can still generate significant conflict. Besides having a target on your back, the Forecasting Department has figuratively built a fort and raised a flag. Blaming everything on the Forecasting Department is too easy, and that can make attacks persistent. It becomes one more department to unload shots on in the corporate circular firing squad.

But residing inside the protection of a fort is preferable to roaming out in the wilderness among your predators. Therefore, a separate forecasting department makes much sense in large corporations that require many detailed forecasts.

The remaining 86% of companies must rely on issues related to corporate structure, company needs, and individual factors to determine the forecasting function's location.

Here is my evaluation of the department options from worst to best.

Customer Service

The bias here is just too much. The department is laser-focused on serving the customer, as it should be. But this can result in ridiculously inflated forecasts. "Remember last October, when Customer X had that huge month, and the phones rang off the hook, and we got yelled at? Better to be safe than sorry."

Finance/Accounting

The people here tend to be narrowly focused on internal factors and the bottom line. This approach can work if the person in the role functions as a forecaster rather than just an accountant. However, to ensure independence, the task needs to be isolated as much as possible from departmental influences.

Sales

Salespeople, of course, are interested in more sales. Forecasting is typically viewed as an annoying task and not given the attention it deserves. Sales personnel also tend not to be very analytical. Depending on the situation, there are high and low biases that can impact accuracy. It is difficult to know all the biases involved if you tally the salespeople's forecasts into a composite number. Typically, your forecast accuracy will vary widely from month to month.

However, salespeople are closest to the customer, and where the action is, so their knowledge is valuable. This can work in situations where customers forecast accurately, and you are accumulating their forecasts to produce your forecast (pure bottom-up forecasting). It is essential to have salespeople involved in the forecasting process, especially for industrial firms.

Operations/Production

Forecasting can function effectively in the manufacturing domain. This department is the primary user and has the most vested interest in having accurate forecasts. Yes, there are biases, but they are the ones dealing with those biases.

The problem here is groupthink. The same department is first developing a demand forecast and then using that forecast to generate a production plan. This introduces confirmation bias into the process and limits critical discussion. Operations/Production is far removed from the frontlines and is susceptible to operating in an isolated forecasting bubble.

Therefore, it is vitally essential for the forecaster in Operations/Production to develop an information network (Chapters 9 & 10) to incorporate outside data and information into the forecast. Monthly meetings within the department also need to be held before creating the production plan. The forecaster should not just send the spreadsheet to the planner without some debate.

Logistics

This is a good place for the forecasting function if distribution is the crucial element for the company. In this case, logistics understands the environment and market dynamics best and is most qualified to do the forecast. The existing biases will be pertinent to the industry and company and can be handled as part of the forecasting process. A forecaster in this department should be able to operate somewhat independently.

Marketing

Marketing is an excellent home for forecasting because it eliminates most biases and allows for the greatest amount of independence. Speaking of bias, this recommendation contains a tremendous amount of bias since I worked in marketing and led the forecasting process at a previous company.

Yes, there can still be bias based on expectations of advertising and marketing campaigns, but typically, forecasting operates as a separate function within the department. The Marketing Cheese may choose to increase your forecast due to optimism of the new ad campaign, but then, who better to forecast the impact of marketing activities than the marketing department? The potential bias is limited to marketing-based assumptions, which *should* be more realistic than for other departments.

Likewise, the bias on new product forecasts should be somewhat contained. New product forecasts are the most difficult. It is assumed that multiple meetings between the product manager and operations will be needed to determine the required production capacity. A reasonable, agreed-upon forecast can be developed through those discussions.

The marketing department is aware of market conditions and trends. It should be well-connected to the sales department and have access to the key sales reports. A forecaster located in the Marketing Department has access to marketing, sales, and customer service reports and personnel but can still operate independently, with minimal bias.

Many marketing people lack the analytical skills needed to forecast, so finding the right person can be challenging. However, it is possible to take someone from finance or accounting, transfer them into the marketing department as lead forecaster, and train them in the marketing/sales aspects.

Another issue is that the Marketing Cheese may not be skilled in analytics and may not understand the problems when a forecast goes awry. They may also not be that supportive if they consider forecasting a political burden on their department.

Independence and Support

Forecasting can function effectively in any department if it functions independently. Independence is essential to eliminating the external biases that reduce forecast accuracy.

Also, it is essential to have the support of your boss. Because, as previously detailed, you will be subjected to attack from various directions, and occasionally you will screw something up. Your boss needs to understand and support the forecasting process you develop. They need to have your back when you stand your ground.

Just Win, Baby!

Regardless of where the forecasting process dwells, the goal is to produce the most accurate forecasts, not the most corporate/politically correct ones. You can and will need to measure forecast accuracy (Chapter 13), but you can't measure corporate political biases.

It doesn't matter what department does the forecast as long as the output is consistently accurate and the process constantly improves. The bottom line, to quote the late Al Davis, former owner of the Oakland Raiders: "Just win, baby!"

Relevant Story #2

Giving math lessons ...

It can be difficult when your job is analytical, and your boss is numerically challenged.

I was meeting with one boss and handed him some numbers he had asked for. He frowned at the paper and said, "These numbers don't look right! How did you calculate them?" I explained how I had done the calculations. I used a shortcut in calculating the routine numbers. He then said, "You can't do it that way. Bring me the data, and I will show you how to do it."

I returned with the data, and he said, "Okay, for the first one, you got 47.9. I'll show you the right way to do it." I just waited for that special moment as he plugged the numbers into his calculator and hit enter. He got a confused look and blurted out, "It's 47.9. How is that possible?"

I so much wanted to say, "It's math." Instead, I took a deep breath and responded, "There's more than one way to get there."

CHAPTER 5

Becoming A Master Forecaster Who Stands Your Ground

To become a Master Forecaster, you must strive to attain the following:

1. Provide the most **accurate** forecasts possible under the circumstances.
2. **Communicate** the market conditions, forecast challenges, and results throughout the company.
3. Become a **Competent**, **Credible**, **Confident** forecaster who is in **Control** of the forecasting process.

Accuracy

You are only a good forecaster if you are consistently accurate. And if you are often inaccurate, you will not be a forecaster much longer. A primary goal of this book is to help you develop a forecasting process that will produce more accurate forecasts. I'm not going to instruct you on how to get to the actual numbers because all business situations are different, and all forecasters vary in their statistical aptitude and approach.

There are dozens of good technical forecasting books explaining the advantages of exponential smoothing versus Box-Jenkins. An excellent statistical book and software package are valuable tools, but they are not the focus of this book. I want to help you with all the other parts of your job – the "soft" skills, if you will. So, this "forecasting" book contains only a couple of basic formulas.

However, no matter your situation or what statistical method is used, forecast accuracy can be improved by developing a solid process and then continuously improving that process. The number one goal of a forecaster is always accuracy. And it matters little *how* you get there; you just need to arrive there consistently. Each process will be different due to company/industry and product variables. But, "Just win, Baby" – by delivering consistently accurate forecasts.

Unfortunately, Accuracy Is Not Always Enough

When a forecast is perceived to be *off*, it will be described as *wrong* or *bad*. You will seldom hear any positive comments unless you publicize your good performance because people expect the forecast to be accurate every time. This is why you need a process to continuously improve your forecast accuracy.

As a reminder:

The forecast is never wrong; it varies in its degree of accuracy.

When you mention accuracy, people immediately ask, "But what is accuracy?" In Chapter 8, we will go through the steps needed to establish an accuracy standard so that you can give The Cheese, or anyone else, an exact answer.

For example, at my final employer, the accuracy standard on the commercial equipment production forecasts was to be within 10% one year out, within 15% for two years, and 20% in three years. We assumed our customers could make reasonable planning and financial forecasts within these standards.

You need to establish the accuracy standard and be able to explain and defend it because if you don't, somebody with the power to fire you may set the standard for you. You may be doing a good job forecasting at 12% accuracy, but your new boss was accustomed to seeing 7% in his previous industry, which is less volatile than yours. You set the standard, then refine your process to hit the goal consistently.

Communication

Communicating is essential to becoming a Master Forecaster/Superior Business Predictor. Once you have a process that enables you to produce consistently accurate forecasts, you must communicate the basis for your forecast and the results. A collaborative forecast approach, by definition, means creating a communication network.

Even though it sounds counterproductive based on the dangerous environment described previously, it is better to overcommunicate than under-communicate. Most stressful situations arise when your boss or The Cheese gets upset over unexpected news. The Cheese hates surprises. So, the forecasting process needs to reduce these flare-ups through steady communication.

If you hunker down in the corner cubicle, faithfully delivering your forecast on time, your value will be invisible to management. You will only be noticed when problems arise and you get blamed, rightfully or not, for the mess. Remember, you have that target on your back, and if you don't move around and defend yourself, you are a sitting duck.

I want you to start flapping your wings and making some noise! I want you to increase your visibility so that your internal customers and The Cheese understand your value to the organization. I know that generating the added attention will be uncomfortable for introverts, but it is necessary and easier than you think.

We will communicate what we learn about the market environment and how it impacts the business and, ultimately, the forecast to all interested parties – even your customers – if appropriate. We are going to communicate the forecast and the assumptions throughout the organization. We want The Cheese to know who you are as a valuable member of the organization and to approach you with relevant questions on the industry and company.

A successful Divisional Cheese once told me his "No Surprises" philosophy. He wanted to be the first person to report any problems with his division directly to his corporate boss so The Cheese would not get blindsided or surprised by hearing the bad news from someone else.

The 4-C Forecaster

To become a Master Forecaster, you must be a 4-C Forecaster. By applying the concepts in this book, you can become:

Competent – Your forecasts will be considered consistently accurate and useable by your internal customers. Your forecasting process is thorough and effective. Your process, and thus your forecast accuracy, improves over time. Your boss and other relevant co-workers will view you as proficient in your job and will come to rely on you.

Credible—By consistently delivering accurate forecasts and keeping the organization informed about the state of the industry, you will be considered credible by your co-workers and The Cheese. Credible forecasts are challenged less often because you support them with solid research. Credibility provides some political capital when the forecast is not within the standard or when you are attacked.

Confident – Once you are competent and credible, you can confidently present, support, and defend your forecast and assumptions. This does not mean you are arrogant, although you do need to publicize your victories. When you are confident, you do not need to be defensive when your forecast is challenged; you can professionally discuss things without a heated argument. Your presentations will be delivered with authority and assurance.

Control – You are in control of the forecasting process. You embrace it and own it. Ownership implies responsibility, and you take responsibility for the inputs and outputs of the forecast. It's your forecast, so you want to be in control as much as possible from beginning to end.

Accuracy – Communication – The 4 Cs

If you provide accurate forecasts, communicate your analysis and reasoning, and attain the 4-Cs, you will meet all the goals of a Master Forecaster who stands your ground. Getting you there is the goal of this book.

Relevant Story

A great forecast ... almost an awful outcome

(The following is the most alarming yet enlightening story in the book. It seems unbelievable, but I assure you that it is 100% true.)

In early June, my boss assigned me to develop a unit forecast by product family for the last seven months of the year. Accounting sent me a spreadsheet, and I filled in the units by month per product family. I also needed to revise the expected prices. When completed, it was our financial forecast for the year, including sales, and profits, which would be sent to corporate.

Now, Accounting could have easily accomplished this task. It was their spreadsheet, the same one they used to report monthly sales and profits. But the Accounting Cheese recognized this task as a turd and promptly passed it on to my boss since forecasting was now a marketing function. And my boss dutifully passed the turd on to me. Yes, now I had a turd in my pocket and no one to give it to.

This forecast was much more important than the production forecasts I was pencil-whipping for Manufacturing. I carefully studied the trends and collected some outside information and data.

I finished the forecast and printed two copies on 8 x 14 inch, legal-sized paper. My boss motioned for me to put his copy in the top tray and that we would discuss it before he gave it to the Big Cheese.

However, during the next week, every time I was in his office, I noticed my forecast was hanging over the tray in the exact same position as before. He had not even looked at it.

What I assumed happened next was the Big Cheese asked for the forecast, and my boss grabbed the spreadsheet out of his tray and said, "Here it is!" And shazam! – my boss had just thrown me under the forecasting bus.

In the movie *Jerry Maguire*, a guy invites Maguire to lunch at a crowded restaurant to deliver bad news "so there won't be a scene." My boss took me to lunch at his country club to inform me that my forecast was so bad that the Big Cheese was considering firing me.

When I ask what was wrong with the forecast, my questions are brushed aside, and I am told I must do better if I want to keep my job. My boss was evidently told to chastise me, and he did that. I doubt that he had ever looked at the forecast, even then.

But now, my job was endangered due to what my boss considered a minor part of my job, and I didn't even know if my forecast was much too high or too low.

I was naturally upset when we returned to the office. I grabbed my copy of the forecast and thought about ripping it in half. Instead, I threw it in my junk drawer in case it got brought up later – like if I got fired.

It was next February when I opened that drawer to store something else when I saw the copy of that forecast. I took it out and compared it to the actual year-end numbers. I checked the actuals against my forecast and could not believe my calculations. My sales forecast was off by 1.9%. Since profit was significant to corporate, I compared that also. It was off by 1.7%. It was the best forecast I ever made in my career. That's right, my highly accurate forecast almost got me fired.

Ironically, I doubt if my original forecast was ever sent to corporate. The Cheese probably sent my *horrendous* forecast back to Accounting to fix it before it went to corporate.

I have told this story many times over the span of my career. People often ask if I marched into the Big Cheese's office and showed him how wrong he was. No, because that's not how things work in Cheeseville. There, that would be known as a career *death march*.

For the Business Predictors

In the analytics world of Business Predictors, we will replace the word "accuracy" with "correct" because, in most cases, accuracy cannot be precisely measured. So, the goal is for your analysis, stock pick, recommendation, etc., to be deemed correct. You will excel in your job if you are correct a high percentage of the time.

Therefore, we must develop a process that increases your "win" percentage and look for ways to continuously improve your score. Consider what your current process looks like and how you might measure success so that when we get to those chapters, you can adapt the principles to your specific situation.

Communication is equally essential for Business Predictors. Keep your boss and The Cheese in the loop to prevent those nasty surprises. Showing the depth and quality of your analysis and process builds your reputation and value to the firm.

The **4-Cs** also apply here. As your process improves, you become **Competent**, and your analyses/projections gain **Credibility**. This significantly boosts your **Confidence**. Maintaining **Control** is usually not as essential or possible as in the forecasting realm, but establishing some control of your process helps.

CHAPTER 6
Setting The Foundation For A Winning Process

This chapter details the transition from a dysfunctional forecasting process to an effective one. It is an example of the resources used to accomplish fundamental corporate change and details my story of beginning the creation of an effective forecasting process at a previous employer.

The Process Needs to Change

My company was growing rapidly. New plants were added as successful new products ripped market share from competitors. The surge in sales stressed out employees and systems. I explained to my bewildered coworkers that the complexity within the company had quadrupled; that's why their jobs were now much more demanding.

My job had become so complex that I literally could not do it. The forecasting needs of a growing multi-plant operation were more than pencil-whipping numbers on a paper spreadsheet. My boss had left the company, so there was no longer that resistance to forecasting becoming an expanded marketing function.

My pricing duties were impacted even more. The pricing system I had set up to handle our complicated pricing strategy worked fine for the current products but malfunctioned miserably in pricing the new products. A relatively basic computer programming change was needed to correct this, but corporate IT had given their staff strict orders to work on only Y2K programming that year.

I had been overloaded with so many job responsibilities that I was doing them all poorly. It was too much for me to handle. The low point came when the Big Cheese berated me because the Canadian aftermarket pricing was all messed up. I was not responsible for aftermarket pricing or any Canadian pricing. Regardless, the attacks continued, growing in intensity.

I was almost fired, but instead, I was relieved of my pricing duties. I would now focus on forecasting and marketing/sales analysis, with an occasional marketing research project thrown in. I was now the chief forecaster whether I wanted the responsibility or not. This was fine by me, not that I wanted to be the forecasting guy, but because I got to stay employed. However, I still had little knowledge and experience doing forecasts, and now I was on another *hot seat*.

The Research

Reliable forecasts were now desperately needed. The production planners were scrambling due to the rapid growth of products and manufacturing plants. The Cheese recognized that better forecasting was needed, and I needed to figure this forecasting thing out as fast as possible.

I could spend more time analyzing the production/sales data since this was now my primary duty. The rest of the time, I researched forecasting methods. I did not have a book such as this one, so I read articles, listened to speakers, and talked to experts and forecasters within my industry. Out of all of this, I began to cobble together a process that made sense for my company and industry. It was very successful and served me well for many years, so I am sharing it with you.

Keep in mind that every business is different, and forecasting is not one-type-fits-all. Use the process detailed in this book as a basis and modify it accordingly. The steps I followed are all here. Again, modify them as needed.

The Company Culture Issue

Your company culture will resist some of the forecasting process changes. That's because all change is negative in the short term. Even the best changes your company ever made were opposed by some people initially. Corporate change happens very slowly, if it happens at all.

I'm not an expert on corporate change. You'll have to read another book on that. The good news is that I started building my new process at the beginning of the year. Everyone in the company was focused on their own new challenges, so my stuff was flying under the corporate radar. Manufacturing put up little resistance to my new process because they were pleased with anything that improved the old system. And they supported my efforts to make the necessary changes.

Yet, throughout that first year, many people, including those from manufacturing, asked me many times, "Why are we doing this?" My response was the same – "Because that's what the experts say to do." They had no answer to that, so I just continued to implement my changes.

But please note: Some of the concepts in this book will seem radical to people, including your bosses, heck, even you. But they do work. So, you will need your boss to buy into some of them. You may need to have them read this book or at least copy the key pages if they need more convincing.

For the Business Predictors

You should always seek additional training/seminars to help you fine-tune your process. What gaps in your knowledge base or skill set do you need to fill and improve on? What new technology is available to help you succeed?

What environmental factors are changing? Which are stressing your win percentage? How can you adapt? How can you change/evade a restrictive corporate culture to improve your score?

Again, evaluate your current process and start thinking about making improvements based on these factors.

Time to Begin

The following chapters will provide you with the tools to establish a new process or modify an existing one. If you already have a process, the materials will help you make modifications and plug some holes, which will improve your accuracy, help reduce organizational conflict, and enable you to stand your ground.

CHAPTER 7
Fundamental Forecasting Issues

Forecast Scope or Level

It is recommended that an individual forecaster be limited to managing 6-15 top-level product families. The goal is to get the aggregate numbers correct before the software or production planners break down those numbers into smaller units.

You can only focus and become an expert on so much. If there are several divisions or many product families, you should have multiple forecasters, each assigned to 6-15 product families. Multi-division corporations typically have separate forecasting departments at the divisional level. Your accuracy will suffer if you overload a forecaster with too many product families. The forecast scope needs to be manageable to be effective.

The number of product families forecasted will depend on the importance and size of the families. At a previous employer, 12 natural product families needed forecasts. Therefore, I focused on and forecasted these at the top level. If you have five key product families and ten others, you can get away with forecasting the six groups, the five important ones, and an "All Other" category. When forecasting industry commercial trailer production at my final employer, there were seven major trailer types and an "All Other" category comprising the rest of the specialty trailers. This was a natural division. The eighth largest segment had limited data and was irrelevant to our customers.

Select your 6-15 product families based on what makes sense for your company/division and data output. After forecasting at the product family level, it is recommended that you use software, custom programs, or macros to generate the forecasts for product sub-groups and SKUs. This is usually a function for the product planners; however, it can be a function of Forecasting, especially if the computer-generated output requires adjusting.

Relevant Story #1

I built this excellent forecasting process at a previous employer that significantly increased accuracy and reduced conflict. Unfortunately, the forecasting process at our sister division remained a toxic mess. Over there, they would abuse and berate the forecaster for a couple of years and then axe them for poor performance. They used their forecasters as punching bags, resulting in high turnover.

I was terrified the Corporate Cheese would have me forecast for both divisions (the product lines were related). However, this would have required me to forecast 22-25 product families. It would have been too large of a scope to navigate, which would have reduced the forecast accuracy for both divisional forecasts. Also, I wanted to avoid dealing with the Bad Cheese at the other division who abused forecasters.

A better solution would have been for me to train their forecaster on my effective process. I never suggested this because I was afraid of "Hey, I got a better idea! Why don't we just have you …."

Forecasting Bookings, Shipments, or Orders?

First, definitions:

Bookings: Units booked into the production schedule that are planned to be built within a given time period – usually monthly.

Orders: Units ordered by customers in a given time period that can be booked according to customer requests over a multi-month time period.

Shipments: Units that physically leave your facility destined for customer delivery within a given time period. Once shipped, the customer is usually invoiced.

As a demand forecaster, production typically needs to know how many units the customers will need (request) in a future month. Therefore, you are forecasting future bookings. Your forecast may be that 1,000 units will be demanded in March.

Orders are typically harder to forecast, especially in industries with wide demand swings or huge orders (blanket orders) that may be produced over a year's time. For example, suppose three large orders are received in August, and only a few smaller orders come in September. In that case, your August order total is 1 million units, and September is only 200,000 – you can't forecast that.

You may choose to forecast and track orders privately to determine if your demand forecast is on track. For example, from historical patterns, you estimate that you should get orders for 10,000 units in July to keep your October forecast on track. If you only get 7,000, you are running behind and need a strong August to catch up. Also, order forecasts can be relevant in stable, cyclical industries. Tracking seasonally adjusted orders and making year-over-year comparisons may help.

Sometimes, The Cheese wants to see order forecasts, and of course, you will produce them, but always remind people that orders can widely fluctuate, impacting forecast accuracy.

Shipment forecasts are essential for financial forecasts since this is what actually generates revenue. Theoretically, monthly bookings should be close to shipment totals, but can vary in some industries. Production disruptions and supply-chain issues could cause a larger gap between bookings and shipments.

Therefore, you will want to forecast bookings. Track orders and forecast them if useful. If bookings vary significantly from shipments, understand the reasons why.

Relevant Story #2

How shipments can vary from bookings – and company culture usually trumps change.

I attended a seminar titled "Project Management for Effective Company Change." The seminar presented a method to manage a *change* project at your company.

This motivated me to start the "Dock Project" when I returned. Many of our customers handled their own shipping and would route a truck to pick up their finished products. This resulted in a lag time between when the products were produced and when they actually left the dock. When several customer pick-ups were delayed, this caused a backup of products on the dock. This disrupted shipping operations, exposed the products to the elements, and negatively impacted cash flow.

I formed a team, held meetings, and developed a comprehensive plan to improve the situation significantly. My boss signed off on the changes. However, when the plan was presented to the Customer Service Manager, he proclaimed, "Nah, we don't want to do that! It might upset the customers." So, absolutely nothing was ever done to improve the dock situation, and I never tried to lead another *change* project after failing on this one. Sadly, not only was the cost of the seminar wasted but all the time and effort of the subsequent project meetings to try to fix the problem. Of course, the problem only worsened as the company grew.

Pure Demand Forecasts

Some forecasting experts will endorse the idea that all forecasts should be pure demand forecasts. This means that if there is demand for 1,000 widgets in March, but your company can only build 800, then your forecast should be 1,000. While theoretically correct, this causes issues in measuring forecast accuracy. Your forecast will always be 20% too high if everything goes as planned for an extended time.

A pure demand forecast is helpful for planning purposes and guides the company in determining how much additional production is needed and how much market share is available. A pure demand forecast is more complex than a standard monthly forecast. It involves discussions with sales management and salespeople. There will be confidential talks with customers and potential customers. You will also need economic and industry data to forecast long-term growth in demand. It also requires a forecast for market share.

Though "Total Demand" forecasts are helpful, they should be done in addition to your standard demand forecasts. You cannot measure accuracy against a Total Demand forecast because you don't have an actual number to measure against. Granted, when you are producing at total capacity, the forecast is equal to capacity, and temporarily, it is easy to forecast but irrelevant. In that case, produce two demand forecasts but measure against the standard demand. It is easier to explain a 2% variance in the standard demand forecast than a 20% variance in a total demand forecast.

Supply Forecasts

When the company is producing at total capacity, you should consider your forecast a "Supply Forecast" because you are forecasting what can be supplied versus the total demanded. A total demand forecast assumes that you could sell more products if you could produce them. A supply forecast is a forecast of what you can supply based on manufacturing/supply chain constraints.

Supply Forecasts also occur due to supply chain disruptions, labor strikes, equipment failures, and other similar events. During the supply chain crisis after the pandemic, many companies were forced into forecasting supply. By comparing your Supply Forecast, which is your standard forecast under these situations, with the Total Demand Forecast, you can provide Finance and The Cheese with an estimate of lost revenue due to the output-limiting event. Under extended supply constraints, the two forecasts are essential for mid-range planning and estimating the amount of pent-up demand.

Relevant Story #3

The Great Supply Chain Crisis of 2021-2022

I first heard reports about supply chain shortages at commercial trailer manufacturers in late November 2020 and reported them to our customers on a company webinar in early December.

I was the first person to inform the Wall Street Journal of the serious problems in the industrial supply chain in early 2021. I'm not really that intelligent. The commercial trailer industry is a leading indicator of the general economy, especially the industrial sector (explained later in the book), and my job was to be an expert here.

However, I followed that stunning achievement with one of the worst predictions of my career. I proclaimed in an April webinar, "Yes, there are supply-chain issues in both truck and trailer markets, but I expect the industry to be able to handle them. I don't expect the impact to last more than another four months." This analysis was based on the fact that truck and trailer builders were well-skilled in working around supply chain constraints due to the vast demand swings in the industry. But this clog in the supply chain turned out to be much worse than others.

As the Great Supply Chain Crisis of 2021 intensified, the number of trucks and trailers that could be built fell significantly below demand. Because we were forecasting equipment builds and sales (basically demand), we were forced to forecast supply instead of demand.

This caused all sorts of problems because all of our models, methods, and history were set up to forecast demand rather than supply. Our team would become frustrated when our forecasts were off, and I would have to remind them. "We are the best demand forecasters in the industry, but we are not equipped to forecast supply. Nobody is."

I developed an estimate of pent-up demand, and if you added that to what was being produced, it served as a total demand forecast. That provided a reasonable estimate of the total demand over the next year. The supply chain crisis was so bad that trying to measure demand in the short term was futile.

Note: If I had known the supply chain crisis was going to last almost two years, I would have developed a process to forecast supply. Imperfect, yes, but it would have been useful and unique. Therefore, if you ever experience a supply chain clog as enduring as the last one, consider developing a way to forecast supply.

Forecast Frequency

Forecast frequency is primarily a function of your business environment. Most manufacturers plan production monthly, and the demand forecasts should synchronize accordingly. Generally, forecast frequency should match production or business planning timeframes.

However, under extreme uncertainty, these frequencies may change. The old adage goes:

If you can't forecast accurately, forecast often.

As mentioned, at my final employer, we ran our April 2020 economic forecast five times in 11 days in March 2020, as the pandemic exploded. We probably would have run it several more times if we hadn't had to meet our month-end publication deadlines.

Forecast Horizon

This depends on what your internal customers want. Usually, production wants a minimum of six months of visibility. Other departments may need a 12-month forecast. Finance or corporate will often need a five-year forecast, usually done yearly or every six months. This can be more complex if the product mix is changing or if new products are expected. While it is tempting to just run the numbers through your forecasting software, a discussion with Marketing is needed to gauge product introductions and deletions.

A suitable method here is to run the data through the software to get a 5-year baseline forecast and then adjust it based on Marketing's input on market share, pricing, new products, new markets, competition, etc.

The Process

Chapters 8 – 17 explain how to build an effective forecasting process.

Chapter 8
Establishing an Accuracy Standard

Forecasts must be *accurate*, but how accurate must they be?

An Old Joke:

Here's a classic: Three economists are out hunting ducks. A duck flies by, and the first economist takes a shot, missing the duck by a foot high. The second economist takes his turn, missing the duck by a foot low. The third economist, in a burst of enthusiasm, jumps up and exclaims, "Yes, we got him!"

The joke highlights the nebulous, inexact nature of the dismal science and makes fun of economists' attempts to deflect their "misses." I have used this story, replacing "forecasters" for economists, to begin presentations, so feel free to share it.

However, in forecasting, close is good enough. The old expression, "Close only counts in horseshoes and hand grenades," can be modified as "Close only counts in horseshoes, hand grenades and forecasting." But how close is "close enough"? That is the purpose of this chapter.

Measuring Accuracy

The simplest way to measure accuracy is: (Actual – Forecast/Forecast). If your forecast was 500 units and the actual was 550, you under-forecasted by 10%. If the actual was 470, you over-forecasted by 6%.

Most statistical types will find this measure much too basic. So, use the most sophisticated method that yields the best results for your situation and understanding. I used the simple method above, which was adequate for my needs and limited statistical ability.

The critical thing is that you must MEASURE and compare every month.

From Chapter 2: **Measure** – Make Design/Process Changes – **Measure**

Where Are You Now?

If you are not currently measuring your accuracy, calculate the accuracy numbers for your product families for the previous 12 or more months to establish a baseline. It is beneficial if there is little difference in demand swings amongst your product families. If some product families are stable but others experience spikes and troughs, the accuracy data should indicate that. You may want to run standard deviations on the data for each product family.

Once you know your baseline and demand variability, you can negotiate your forecast accuracy with your internal customers. This is the first step in *team forecasting*. The goal is for you and the forecast users to agree on an accuracy standard. This may be difficult if your current relationship is strained, but it is necessary to build an effective process and should be done regardless. This discussion will improve your relationship with your internal customers in Manufacturing/Materials.

Where Do You Need to Be?

To develop a process based on continuous improvement, we need a target to shoot for. We will measure our progress every month and need to know how well the process is performing to achieve the goal.

It's challenging to set your own target because you do not know what your internal customers, your data users, need or their expectations. However, you do know what an unacceptable forecast looks like because that's when all the howling and bellyaching begins. Also, if you set your own goal in a vacuum, human nature causes you to set goals you can easily hit. And this produces little benefit.

You will employ a simple marketing concept to determine an acceptable forecast accuracy. You are going to ask your internal customers what they need. What a novel concept! Instead of arguing, back-biting, finger-pointing, etc., you simply ask, "What do you need from me to succeed?" or "Help me help you." These powerful statements show that you want that person to succeed in their job, and you are willing to help them do that. That philosophy is sorely missing in corporations today. It is the ultimate teamwork statement in that you want to win together.

The best way to control this part of the process is to visit the department manager, who uses your forecast numbers to plan production. I do not recommend setting up a formal meeting with several people to discuss this. The more people involved, the more likely the meeting will spin out of control. And you definitely don't want any Cheese there, lest it turn into a political game.

Before this discussion, you will have analyzed your forecast history and, know your accuracy history. This also indicates how much improvement is possible. These details will be part of the negotiation.

The other part of the negotiation is to determine the "key forecasting month". The key forecasting month is the month in the six-month forecast that is most critical for production planning purposes. It indicates how quickly the manufacturing/materials function can respond to changes in demand.

For example, measuring against the one-month forecast is meaningless since production can't react that fast. Three months is a common window, but your company's range could be longer depending on the product and the global supply chain.

The standard you choose to present is either based on the potential for improving your past performance or the variability of your product demand. Throw out a suggested standard so that the person knows what you need. I included the 15%, 3-month standard because that is what the materials manager gave me when I asked him at the beginning of my process development. Your standard will be different because your business is different.

The other reason to throw a standard out there is that this is a negotiation. You don't tell them it's a negotiation, but the goal is to agree on a target. You are drawing this target together, increasing the chances of buy-in and cooperation going forward.

Before you talk to production planning, discuss lead times for components with purchasing to see how they manage materials. Let's say a key component manufactured in India has a 5-month lead time. This doesn't mean your key forecast month would be five months. You would expect the item's inventory to be higher due to the longer lead time of the components. So, maybe there are two months of safety stock on the item. Therefore, the key forecast month, which is still dependent on Manufacturing, may still be three months.

If three months is the key forecast month, we will still measure accuracy against all six months of the forecast, but we will focus on the three-month forecast. In practice, the three-month forecast becomes your target's bullseye.

The Negotiation

To stay in control of the negotiation, you will want to present an opening proposal for your manufacturing person to consider. You can't just ask how accurate your forecast needs to be because it lacks the required context, and the answer won't be specific enough.

Let's say that after your research, you determine that you are usually +/- 18% on the 3-month forecast for most product families. Your personal goal is to tighten the range to +/- 15% in the next year based on improvements in your forecast process.

Your simple opening statement is:

"I'm taking steps to improve the forecast process and want your input. What is a good forecasting range in terms of accuracy and length of time that will allow you to plan

production effectively? If I get you a forecast within +/- 15% three months out, would that work well for you?"

The manufacturing manager will likely want to review and discuss his data with his team. Separate accuracy/time standards may be established for different product families if response/lead times differ significantly.

Ideally, they respond with a realistic accuracy standard. If the person is professional and supports your efforts to improve accuracy, it should be reasonable and attainable. "I want you to be within 5%, six months out" is a wish, not a goal (in most cases).

"Just give me a good forecast" is insufficient because it is not a goal and cannot be measured. They should be pleased that you are working to improve the forecast, but if they refuse to help you set a goal, then just set your own.

Let's say the manufacturing manager wants +/- 13% accuracy for the 3-month forecast. You respond by saying that your beginning standard will be +/-15%, with the goal of achieving +/-13% over time.

So perhaps your initial goal is 15% because you will need time to develop and tighten the process. In addition, at some point, you will publish your forecast accuracy (much more about this later), so you need a reasonable goal to start with and then tighten your standard as your process improves.

So, you go back to the manufacturing manager and say:

"I appreciate your help in establishing a forecasting accuracy goal. I want to set the long-term goal at 13% accuracy in the 3-month forecast. As previously mentioned, I am making significant changes to the forecasting process intended to improve accuracy. I will start with a forecasting standard of

+/-15% for the 3-month forecast and then improve the process with the goal of achieving 13%. Once we get to 13%, I may be able to do better than that."

If they agree, you have an accuracy standard and a place to begin. You have also involved Manufacturing in the forecasting process and are building teamwork.

Note: Even if you use more sophisticated methods to measure your accuracy, you should use the basic method when communicating measurements with other departments. Most manufacturing people will not understand mean absolute deviations and don't need to.

Why are you doing this?

This is the first step in improving your forecast process. Maybe not here at the beginning, but indeed, along the way, you will be asked:

"Why are you doing this?"

And just as I used the "experts say so" as a justification, you may say:

"I read this book by a guy who knows a lot about improving forecast accuracy, and he suggests doing it this way."

Yes, throw me under the bus if you must. I've been under the bus so many times in my career that I still have the tread marks! If they want to ask me questions directly, they can contact me at donake@outlook.com.

Now that we have an accuracy target, it's time to start figuring out how to hit it. Establishing a standard and consistently hitting it is a key factor in being able to stand your ground.

For the Business Predictors

It should be standard practice to ask your internal customers what they need to get from your business analysis or research project. "We would like you to study this" doesn't cut it. Neither does "We need to know stuff about our market" for a marketing research project. A good research project has specific questions that the analysis will answer or at least attempt to answer. It is critical for marketing research projects to know what key decisions will be made based on the information collected by the project. If you don't do this, you risk issuing a report with lots of data, charts, and graphs and then suffer the glare when The Cheese is confused about making the decisions.

Ask the questions up front and document the answers because bosses and The Cheese often have conveniently faulty memories. These answers become the project goals and should be stated in the first paragraph of the research report. Don't let them burn you with criticism of a faulty analysis because you studied something other than what they *really* wanted. The misunderstood communication always falls back on you.

Here's A Trick

You will often be involved in rambling business discussions where the group needs help narrowing the focus and developing conclusions. When asked to give your opinion, reply, "Please ask me a specific question." This may be an effective way to determine the precise goals and to steer the discussion into the essential factors.

Chapter 9

Building Your Information Network

A key to becoming a Master Forecaster/Superior Business Predictor is having information no one else does. To do this, you need to build a business network - not to advance your career, but to provide you with this valuable intel. These *golden nuggets* will enable you to generate accurate forecasts and correct conclusions/recommendations.

The goal is to build a network of sources, inside and outside the company, that can provide you with vast sources of information, both qualitative and quantitative, with which you can support your forecast and analysis conclusions/recommendations.

Building this network takes work; however, once constructed, it is a potent tool. It enables you to be one of the most knowledgeable people in your organization and even your industry. It increases your value and status inside your company. It also builds your reputation in your industry, which helps if you ever need to change companies.

Some of the most valuable information must be obtained by phone calls rather than text or emails. You will need the opportunity for questions and comments back and forth, and some semi-confidential information (This didn't come from me) is best not typed. I realize this will be a challenge for the introverts and younger forecasters, but it is not difficult to accomplish and is well worth the effort.

Internal Information Network

Marketing Department

Establish a contact in Marketing and communicate regularly. Information from Marketing is essential for consumer products but less so for industrial ones. Make sure your marketing person knows to inform you when new information pops up.

Relevant information from Marketing:

- What promotions are planned (during the key forecast month), and what is the expected sales increase from these promotions?
- What price changes are planned during the key forecast month and why? What is the expected sales impact of these changes?
- What is the competitive environment? What product families are gaining/losing market share?

New product forecasts should be handled separately from the standard forecasts. Marketing and production should hold separate meetings until the new product demand falls into a more traditional trend and can be incorporated into the regular forecasting process.

Remember, the information you get from marketing will be biased. They will tend to overstate their expected sales from the promotions, so you need to double-check the impact of past promotions. They will often magnify the gains of cutting prices and understate the hit of price increases.

Promotions complicate the forecast because production may need to increase before the promotion period to provide the inventory required to support a sales spike. This should be discussed at the Sales and Operations Planning/Forecasting Meeting (covered later). If new promotions pop up, Manufacturing should be told immediately.

You should also check with Marketing when developing longer-term forecasts. The marketing plans may include new product introductions, product phaseouts, new product branding, new promotions/pricing, etc., that could impact demand in the next 12-24 months. Remember, marketing plans tend to be overly optimistic by nature. However, The Cheese usually wants you to use the marketing plan numbers in your long-term forecast.

Bottom line: Determine the person in marketing who knows what's going on, and check in with them every month.

Sales Department

Sales has access to valuable customer information relevant to your forecast. Of course, salespeople are paid to be focused on making sales and often forget to report the news you need. You won't have time to talk to every salesperson, so you must figure out the most efficient way to extract the information.

Who you check with monthly depends on the size and scope of the salesforce. If it is relatively small, you might communicate with a couple of salespeople and the sales manager. If the sales team is huge, you have to find the right managers who have the information relevant to the forecast. Regardless, salespeople and sales managers need to know that they are part of the forecasting process. They are responsible for reporting news that can impact the forecast. They have a vested interest in the process because an accurate forecast that is managed well leads to increased customer satisfaction and, therefore, more sales.

You will have established an effective sales communication network when the following types of information are reported to you without you having to ask:

- Customer A needs 40% more product in April ahead of a plant shutdown in May.
- Customer B needs 30% more products in June, as they are opening a foreign warehouse.
- Customer C is planning to cancel a large order in May.
- Customer D's sales are expected to decline due to competitive pressures.

An ineffective network results in systems failure, conflict, and the blame game. Of course, sometimes salespeople just forget to tell you critical news, which can result in an inaccurate forecast.

It is always difficult when you are expected to know something important but don't. Blaming sales for the lack of communication sounds defensive and lame. The best way to handle this is to say, "No one reported that to me."

But then, the next step is to go back to the person who had that information and explain what happened with the forecast and the heat you just took from whomever. Don't blame them for anything, but urge them to feed you any pertinent information in the future. It is counter-productive to rip into them just because someone ripped into you.

Even if you routinely talk to the sales managers, it can be enlightening to occasionally speak directly with salespeople.

Every sales team has a couple *of big-picture* thinkers who can offer unique business perspectives from the front lines. Find out who they are, have lunch with them, and build those relationships.

Regardless of what sales personnel you contact, here are the key questions to ask:

- Are there any new quotes this month, especially those hitting in the key forecast month?
- Are there any existing quotes that could close in the key forecast month? Are there existing quotes that once looked promising but have cooled off?
- Are there any previously booked orders that may be moved up, moved back, or canceled?
- Where do you see sales going in the next six months, and why?
- What market trends do you see happening?
- What are customers telling you about the market?
- What is the competition doing to thwart your efforts?

You will listen intently, take detailed notes, and thank them for their time.

Relevant Story #1

Yes, somebody screwed up – and it was you ...

One time, I took considerable heat because of an under-forecast that resulted in a missed delivery date to an important customer. A large order was entered late that wasn't accounted for in the forecast or production plan.

Before I could research what happened, the salesman called me and chewed me out over the error. "Didn't you know about this critical order?" he exclaimed. I responded that I did not. Of course, I didn't because that very salesperson had failed to tell me. I took a deep breath, and we continued the conversation. At the end of the call, I suggested that in the future, it would be helpful if he informed me of those types of critical orders so we might be better prepared. And he got the message because communication improved after that.

The Internal Experts

Every company has a few intelligent employees, *big thinkers* who strive to understand the business on a higher level. Some of these grizzled veterans have been with the company for many years and have experienced several business cycles and environments. These people can work in any department; they may not even be in management. One of the sharpest people I worked with was a quality control technician. He read The Wall Street Journal and understood the macroeconomic forces impacting the business. Our conversations were always valuable.

Identifying these individuals and opening a line of communication with them is beneficial. Ask them their opinions. When you have conflicting data or information, bounce it off them. You will gain valuable insight and win their support, which is helpful when the attacks come and you need to stand your ground.

The External Expert Network

Building a network of external experts takes time and considerable work. However, it is worth the effort because the information it provides is critical to becoming a Master Forecaster or Superior Business Predictor. It will give you information no one else has. It provides various pieces of the puzzle, which allow you to form a picture of where your market or sector is going.

Creating the Network

First, you will identify the people in your industry with knowledge and information relevant to your forecast. They can be customers, suppliers, industry colleagues, or financial analysts. You find the names from your salesforce, industry press, trade/industry meetings, coworker referrals, etc. This is an ongoing process as people exit and enter the industry, and you are continually refining the network to include the best people available.

Next, I recommend cold-calling these people for the initial conversation if you have never met them. Some people tend to ignore emails from people they don't know but are more likely to return a voicemail. Email is preferred when initially contacting a *status* person such as a national economist. Most of these people won't talk with you, but it is a coup when they do!

Regardless of who you are talking to, the first call starts like this: *"Hi, I'm Joe Guru from XYZ Corp. I'm in charge of forecasting here and trying to figure out this market. Can I ask you a few questions about what's happening in the industry?*

Oh, and anything we discuss here will be kept strictly confidential."

Most people will be flattered that you are asking their opinion and will proceed with the conversation. Limit the discussion to three questions and only take up 15 minutes of their valuable time unless the discussion is going so well that they keep talking.

A pivotal moment comes after they finish answering the first question. At that point, you will give them your opinion about the subject. It is vital that you provide them with something in return because:

YOU ARE TRADING INFORMATION

You are getting information, and you are giving them back information. If the person doesn't get anything out of the call, you will have taken up 15 minutes of their time, solely for your benefit. This makes it unlikely the person will talk to you again.

You will need to have studied the topics in your three questions before the call and be ready to provide your insight. You are not interviewing the person but engaging them in a mutually beneficial dialogue.

You can ask follow-up questions, but limit the call to around 15 minutes or so. Thank the person for their input, and suggest you will check in again in a few months. Add their name and number to your expert list spreadsheet and record when you spoke with them. Review your notes and highlight the key points. You will want to report essential information in your forecast assumptions report and the S&OP/forecasting meeting.

Of course, in building your expert network, you will inevitably contact someone who is not helpful or knowledgeable. You get answers like "Gee, I don't know" to your questions. In those calls, you will end up giving them more information than they gave you. Here, you end the call by thanking them for their time but not saying anything about future calls - because there won't be any. People have to earn their way onto your list. You are seeking intelligent, helpful experts, and some people will not make the cut.

Try to add a new contact to your expert list every month and, of course, more at the beginning of the process. Do not contact anyone more often than every three months. As part of your forecasting process, you will communicate with three to four contacts monthly. For business predictors, check in with outside and inside experts as part of your process. Market researchers may want to consult with some external sources before designing your questionnaire.

Confidentiality and Trust

During discussions with your external expert network, confidential information - the information they are willing to tell you but don't want to have repeated - will come up. It is important to emphasize that you will keep that information private. I would even further assure the person that I would not discuss the information within the company and would only use it in formulating my forecasting assumptions.

Over time, the goal is to build trust within your network. Once people trust you, they will feed you information that they will not share with anyone else. They may even reveal things they shouldn't be telling you. Sometimes, I would even preface a question with: "I don't want you telling me anything proprietary, but could you answer this question for me?"

I would keep the source of this type of information confidential within my own company because I did not want someone accidentally revealing my source to someone outside the company.

Sometimes, at my final employer, I would report interesting information in a meeting, and my boss would respond with: "Where did you hear that?" If it was proprietary, I said: "You don't need to know." Of course, it helped that my boss trusted me to report accurate information and respected my ability to extract essential market information from my network.

In my career, I never violated trust or blabbed about confidential information. And as an industry expert with great connections, customers, reporters, and financial analysts often asked me to spill the beans.

One time, I was accused of leaking confidential information by a large customer. The funny thing is the information was not that important, was not highly confidential, and could have been blabbed by a number of people, including delivery drivers servicing their plants.

But, The Cheese at this company were crazy-paranoid, and because I was considered an industry expert who knew lots of stuff, they assumed I must have been the leak. I vehemently denied the accusation. I did know about the information, but, of course, I couldn't prove that I didn't reveal it. Fortunately, my Cheese believed me, and the whole controversy faded away.

The Benefits of Expert Networks

By networking with industry experts, people will begin to consider you an industry expert - because you are! You will have information no one else has because you went out and got it. This gives you an advantage when making and supporting your forecast or developing your business analysis. And because you are an expert, people will begin to contact you for discussions, which further enhances your network.

Being an industry expert also builds your value and reputation within your company. You know you have accomplished something when The Cheese approaches you with industry-related questions. Having a powerful expert network is vital to being a Master Forecaster or superior Business Predictor. Having exclusive, relevant information helps you stand your ground.

Relevant Story #2

Expanding the network

You do not need to limit the expert contacts to your industry. If you find an intelligent person with relevant information, add them to your network.

One of the most intelligent members of my network was an economist and analyst in the railroad industry. I met him at an industry conference and would check with him occasionally to discuss the macroeconomy related to commercial freight.

One time, my analysis indicated the economy was in recession, but almost no economists had stated this. When I called the guy to discuss this, he replied confidently, "We've been in a recession for a few weeks. It started last month." And he was correct!

Relevant Story #3

Don't gab about proprietary information – especially to your competition

One time, my boss blabbed, actually bragged, at a trade show to a competitor about a major product research project we were funding at a large university.

The competitor then called the university and spoke to the project leader. And apparently, the scientist, being proud of his research, freely shared the general results and conclusions of the study with them.

You can imagine my surprise when a trade journal advertisement from our competitor referenced the research findings, mentioning the university, but of course, not our name. We had yet to even receive the final research report! Everything in the ad was accurate, and it was our research, so we had absolutely no recourse.

I only learned the back story when I showed my boss a copy of the ad.

Moral of the story: KEEP YOUR MOUTH SHUT!

Chapter 10

Building Your Data Collection

Chapter 9 involved collecting qualitative data. Now, we focus on collecting and organizing quantitative data.

Internal Data Analysis

Every month, you should review the following:
- Charts showing the shipments for each product family.
- The forecast accuracy charts (Covered later).

You can run various statistical analyses on the data and, if helpful, seasonally adjust the data. Quantitative data analysis is beyond the scope of this book. Still, you should try to find the analyses relevant to your situation and run them monthly, early in your process.

The forecast accuracy graphs show how much you over/under-forecasted each month. It helps identify your forecast bias. If you over-forecasted the last four months, your forecast assumptions (covered later) have been too optimistic. Knowing your biases helps you in the final forecast adjustment. If your forecast looks aggressive and you have over-forecasted the previous four months, you want to cut back some.

Every week, you should review the following:
- Weekly orders (Macro, not per product family)
- Weekly shipments (Macro, not per product family)

You will want to track the orders going into your critical forecasting month to see if the backlog for the month is ahead or behind your forecast. You can set up guideposts based on history to measure this. For example, historically, the key forecast month is 50% booked at your first measuring point. If it is only 45% booked, bookings will need to catch up. It is basically a measurement of the quality of your latest forecast. This is vital information to have when evaluating the current and future direction of demand.

Tracking weekly shipments indicates how the current month is trending. It gives you the first look at how your forecast accuracy may turn out. A Master Forecaster should always know where the market is headed. In addition, a change in market direction has implications for your next forecast.

At a previous employer, this information was reported daily, and I would enter the order and shipment data into my tracking spreadsheets. This allowed me to keep my finger on the pulse of the market and answer questions if The Cheese happened to stop by.

I recommend doing the internal data analysis early in the month since it gives insight into where the market and forecasts are going and enables you to develop hypotheses. These factors and scenarios are good things to discuss with the internal/external network to gain essential insight.

For example:

"My analysis shows some fundamental weaknesses in Product E. What are you seeing, and why do you think that is?"

External Data Analysis

Economic Indicators

Thousands of economic indicators are available to you; the challenge is to find those relevant to your business. Of course, this can be difficult because your company has no magical leading indicator. And the most relevant ones can vary in importance as the economic and business cycles change.

Back in the 1990s, there was even a company that would take your sales data and match it up with the best leading indicator it could find. However, this introduces the correlation versus causation factor. This company might find a leading indicator with a high correlation to your sales but no logical causation. Try telling The Cheese you are basing your forecast for refrigerators on demand for house paint. But it still could be a good indicator. (More on this later in the chapter.)

Some books are available that help you understand, choose, and analyze economic indicators. Most are geared toward using indicators to predict the stock market and make sound investment decisions. A good book on this is "The Secret of Economic Indicators—Third Edition" by Bernard Baumohl.

The reliability and consistency of many economic indicators have diminished over the past decade. The Great Recession (2007-2008) knocked many indicators off base for several years. The indicators data were regaining some consistency when the COVID pandemic jumbled the numbers again. Interestingly, few books on economic indicators have been written in the last decade. Economic indicators are still relevant; they are just harder to analyze.

The challenge is identifying the economic indicators that correlate to your sales and have a reasonable degree of causation. Of course, leading indicators are the most valuable, especially if credible, free forecasts exist for them. If indicators have a high correlation to your sales, it may be beneficial to pay for these forecasts. Coincident indicators (happening in unison) can also be helpful if those credible forecasts exist.

Use logic to select/search for the best indicators; housing starts for building materials, for example, and then test for statistical correlation. Once you identify your key indicators, track them monthly. Do not forecast directly from the indicators, but use them to support and check your forecast. If all your key indicators are going down but your forecast is going up, the disconnect must be investigated. The indicators will be detailed in your Forecast Assumptions Report (Chapter 11). You will also show the relevant indicators in any presentations.

Coworkers will suggest other indicators that may be relevant to the analysis. If you have looked at the indicator in the past, you can respond that you have considered it but didn't find a correlation. Nonetheless, it would be best to look at it again because conditions change. When new indicators are suggested, you should review them, especially when mentioned by The Cheese.

If you uncover an economic indicator that highly correlates to your sales but doesn't make logical sense, track it, but don't publicly reference it. There may be a causation factor deep in the data that you can't determine. You lose credibility if you can't explain the logic behind your key indicators.

Understanding the data and indicators that support the forecast is essential in helping you stand your ground.

Relevant Story #1

When asked to do the impossible, I sometimes fail ...

It was a typical Monday morning when my boss burst into my office, out of breath from running up the stairs.

"We were discussing the forecast in the staff meeting, and we figured out that all we have to do is find the leading indicator for our business and do the forecast off of that," he exclaimed.

This gives insight into how easy The Cheese thinks your job is. Just find the magic bullet, and the forecast does itself.

I was stunned that they had discussed this and thought finding a single leading indicator was that simple. My boss stood there like an eager dog who had just found a juicy bone. I held my breath, and then...

"And what did you tell them?" I asked.

"I told them you would do it!" he proudly proclaimed.

And with that, he spun around and dashed off just as quickly as he entered.

As previously mentioned, working in a demanding environment can be frustrating. However, these challenging, even impossible, tasks push you beyond your boundaries and make you better than you ever thought you could be. In this case, what doesn't kill you makes you smarter.

At the moment, I thought: *Crap, these goofballs believe there is actually one leading indicator for our business out there, and they think I can easily find it if I search.*

But I couldn't ignore the command. My boss had enthusiastically told The Cheeses I would find this Holy Grail, so I proceeded.

The task at hand was to find the leading indicator for commercial trailer demand in the U.S. So, I started by examining the litany of obvious, logical economic indicators available. Then, I went to the next level of not-so-obvious indicators where a correlation could exist.

The good news was that many of the indicators correlated with commercial trailers. The bad news was that they were all lagging indicators, happening after the change in trailer demand. I had evaluated all the logical options and failed to find even one leading indicator for the business.

There is an old poker adage that if you can't identify the weakest player at the table, you are that weakest player. In this case, if I couldn't identify the leading indicator, then my sector was **the leading indicator,** for the entire goods-based macro-economy, no less. I would have liked to have seen my face when this epiphany hit after months of work.

While this was a nice theory, what was the logic that backed it up? At the time, the "hot" leading economic indicator was the production of cardboard boxes. The logic is that demand for cardboard boxes preceded the overall market for goods soon shipped in those boxes. The popularity of using this data as a leading indicator was such that an industry group collected and sold the data on cardboard box production.

It is a great leading indicator that is still used today. Over the years, forecasters using this data have no doubt had to adjust their models to account for the millions of home deliveries we have now. As I write this, an article reports that the demand for cardboard boxes is dropping at rates similar to the Great Recession. However, this is more of a function of reduced online sales than an economic downturn.

But after these boxes are packed with goods, where do they go next? (In the early 2000s, before the Amazon effect). Well, most of them are put on commercial trailers and hauled somewhere. What is the lead time for producing cardboard boxes? I don't work in that industry, but I estimate 2-3 weeks. From order to delivery, the lead time on a commercial trailer is about four months.

Which is a better "leading" indicator for the "goods" sector of the economy? I contend it is commercial trailer demand. Cardboard boxes carry primarily consumer goods. In contrast, in addition to consumer goods, the various trailer types haul heavy industrial products, construction materials, and commodities, thus involving all sectors of the goods-producing economy.

What about reporting my findings back to The Cheese? Of course, like many wild corporate goose chases, I was never asked if I had found the leading indicator for our business. This project could have been considered a colossal waste of time; except I learned so much through the process. Examining which economic indicators impact your business is well worth the effort. Knowledge of and reporting these indicators is part of becoming an industry expert and Master Forecaster/ Superior Business Predictor.

In addition, as the following story explains, I discovered that commercial trailer demand was a leading indicator of the stock market at the time.

An Addendum – Relevant Story

As previously mentioned, the commercial trailer market is a leading indicator of the general economy. My curiosity led me to compare trailer production to the S&P 500. I found a correlation. A downturn in trailer production usually preceded a drop in the stock market by about 8-9 months. I was not confident enough in the model to sell any stock before the 2000-2002 plunge.

I continued to track and refine the model. At an industry conference in May 2008, I told several financial analysts that my model predicted a drop in the stock market in 4-5 months. Of course, they were highly amused by this. I wasn't a stock market analyst. Heck, I wasn't even a statistician. I was just this guy who forecasted demand for a trailer component.

In early September 2008, I sold a chunk of my stock holdings because my model was flashing red. My stock broker's reaction was, "Why are you doing this?" After my explanation, he said, "Okay, but I can't support this move." And a few weeks later, Boom! The stock market sank.

Over the next few weeks, I got calls from a few financial guys who remembered my ridiculous prediction at the conference. After the pleasantries, I was asked, "Don, could you explain more about the model?" I would then tell them the logic, including the cardboard box factor covered previously and the historical analysis.

Is my model still relevant? Who knows? The Great Recession (2008-2009) threw off most indicators and the historical economic trends for years, and even though trailer production cycled some, the stock market kept growing. Just about when the economy and the trailer industry returned to some degree of normalcy, the pandemic hit, followed by the subsequent supply chain crisis, throwing the economic indicators and the trailer market into disarray for years. Maybe someday, things will get in balance so that my model can be predictive again.

CHAPTER 11
Making And Documenting Forecast Assumptions

The old adage about making assumptions does not apply to forecasting. To generate an accurate demand forecast, you must predict every factor that could impact your forecast. These predictions, or assumptions about the future, become the base on which your forecast is built.

Your task is to paint a picture of your business environment several months in the future (the key forecast month). Since you can't know precisely what this looks like, you must do your best with the information and data you have and make assumptions about what it all means.

Extract the Useful

You have collected a vast amount of information and data. Now, the challenge is to extract the key factors and determine how much impact you think they will have on your forecast.

You take all the notes, articles, and data you have collected and try to combine these pieces to solve the *forecasting puzzle*. In some months, the pieces will fit together nicely, and you will be confident of the picture of the future you create. Other times, the information may be inconsistent, and the picture turns out foggy. It is okay if some of the forecast assumptions you make are contradictory. This shows that the market is volatile and highly uncertain. Conflicting information typically occurs during market transitions.

Relevant Story #1

Busted by the quality police ...

On the day I would develop my monthly assumptions, I had highlighted notes, graphs, and articles spread all over my office so I could combine similar pieces to create my "picture".

Unfortunately, the company had instituted a clean office initiative and audited my office on "assumption day". I actually got written up for having a messy office, which is hilarious because the next day, with the papers for the month all neatly tucked away in a file, I would have passed easily.

Why the Forecast Assumptions Are Critically Important

Remember: The forecast is never wrong but can be inaccurate (outside the standards).

And if the forecast is inaccurate (versus your standard), the forecast is not wrong, but the assumptions were. Your forecast is the product of your assumptions. You do all the work extracting the data and information so that you can make the best assumptions possible. If your assumptions are correct, barring a surprise event, your forecast should also be. But you are still attempting to predict the future; thus, some months, your assumptions will be flat-out wrong, and your forecast accuracy will suffer because of it.

Your assumptions are tightly connected to your forecast. Therefore, you do not want to argue or debate with anyone about the forecast number, but you always want and are willing to discuss forecast assumptions.

There will be a discussion of how to defend your forecast later in the book, but here is an example of why forecast assumptions are a critical part of the forecasting process:

Them: "The forecast was wrong!"

You: "November's forecast was not within our accuracy standards."

Them: "Well, it looks wrong to me!"

You: "Yes, we assumed this (fill in the blank) and that, but those assumptions were in error, causing the forecast to be 19% too high."

You want to avoid arguing about the number, especially in this case, because the forecast was indeed out of standard. You want to discuss the market, especially if the person has information pertinent to your next forecast. Talking about the assumptions turns a useless debate about the numbers into a potentially fruitful discussion about the business. And this is the same type of discussion you want to have when your recently released forecast is challenged, and you must stand your ground.

If you have no forecast assumptions, the conversation becomes:

Them: "The forecast was wrong!"

You: "Yeah, I missed it last month."

Them: "What went wrong?"

You: "Gee, I don't know, but I will try to do better this month."

If this conversation is with The Cheese, you've got a serious problem.

Because the assumptions are so important, you will document them each month. This allows you to keep a record of what the forecast is based on. It is critically important to know why your forecast didn't hit the accuracy target. You will also need to keep all your notes and documents related to your assumptions in case you need to review them later when the actual numbers for that forecast month are released. Someone could have given you bad information, and you may want to follow up.

You can take this one step further if you wish, and attempt to forecast the impact of each assumption on the forecast. You list the assumptions with a positive or negative value assigned to each one. Therefore, if your three positive assumptions add 10,000 units to the current market but your two negative ones cut out 7,000 units, you will increase your forecast by 3,000.

Documenting the Forecast Assumptions

Break your assumptions into Economic, Industry/Market, and Company/Internal. If needed, you may have separate assumptions for different groups of product families. For example, if the environment for Product Families A through F significantly differs from G through K, then list those assumptions separately.

It is important to note that you are making your forecast assumptions for your key forecast month. If your key forecast is three months out, then your assumptions are focused on the market three months out. Of course, you can include factors happening before or after the 3-month window in the report, but you are making assumptions about your key forecast month.

Examples:

Economic

- GDP is expected to slow to 2.4% in Q3, which could weaken the demand for some (listed) of our products.

- Electric and appliance store sales increased for the fourth straight month in June. This should support our sales in relevant product families going into Q3.

- New Home Sales were up only 0.2% in February and are forecast to be flat for the next two quarters, presenting a challenging environment for sales growth.

- The current ISM (Institute for Supply Management) report shows that business inventories are too low. Reports from the field indicate our customers are planning to rebuild their inventories over the next six months.

- The current ISM report indicated that Chemical Products contracted for the second straight month in January. This could be the start of a downward trend.

Industry/Market

- Customers continue to transition from Technology A to Technology B. This means there will be an ongoing shift in demand from Product Families C and D to Product Families G and H.

- Competitor A just signed a deal with retailer Tar-Mart. This is expected to significantly decrease our shelf space and reduce Product Families B, C, E, and F sales to the customer.

- Competitor B has cut pricing in the Eastern Region by 10% to try to gain market share. We are not responding to this right now.

- Competitor C is running a major promotion in Q3. This is expected to decrease Product Families G, H, and I sales by 2%.

- Competitor D has not recovered from the warehouse fire in Dallas. We continue to pick up market share in all product families in the Southwest region.

Company/Internal

- The new deal with Baxter begins in September, increasing demand for several product families.

- We lost the Cornet bid for 40,000 units over the next 12 months to Competitor E.

- We have numerous bids out for large orders for Q3. We have a strong chance of closing two of them.

- Wal-Blue is running a big promotion in Q4. They want a stock inventory of 30,000 units of Product Families E and F to begin arriving in mid-August.

- We are cutting prices by 6% in the Western Region to regain some market share. We expect a 5% increase in sales within four months.

The Forecast Assumptions Report

I highly recommend compiling a Forecast Assumptions Report and distributing it to the "forecasting team" and anyone else who wants a copy.

The report consists of everything relevant, or potentially relevant, you have found in your information/data collection and analysis. It is divided into three categories: Economic, Industry/Market, and Company/Internal. How extensive your report is will depend on your specific industry/company factors, the scope of your information/data network, company interest in the report, and your available time. My report was comprehensive and averaged about six pages long. I recommend using bullet points to allow readers to skim through it quickly.

In the **Economic** section, you will include graphs of the standard economic indicators, those most closely related to your industry. You can also have charts of any of the fringe indicators that appear most relevant right now. Put comments/analyses of the indicators in bullet points and include statements about what economists forecast for the key indicators. You are documenting facts/analyses here. The forecast assumptions based on all the information will be in the report's final section.

The **Industry/Market** section of the report includes relevant bullet points based on your research. You can include comments from your sources here. Do not include proprietary information or source names in the report. Even though you may use the information for your forecast, you don't want it *in print*. You can protect your sources by saying, "Reports from the field indicate...." Or, "A key customer says that business has never been better."

The **Company/Internal** section includes the bullets specific to your situation. Try to be as objective as possible. The statement "Product Family D is losing market share" may result in a visit from the angry product manager. Soften it up with, "Product Family D appears to be struggling to maintain market share."

The Forecast Assumptions Section

After reading the first part of the report, people will understand the total business environment. Then, you extract assumptions from previously stated information, facts, data, and opinions.

The assumptions provide the basis for increasing or decreasing your previous demand forecast. It is easy to determine direction when all the assumptions are negative or positive, and the forecast adjustment becomes that of degree. When dealing with a mixed bag of assumptions, the task becomes much more difficult because you must determine which assumptions are most likely to occur and which have the most impact.

Benefits and Uses for the Forecast Assumptions Report

My most significant personal benefit in writing the report is that it required me to review all my data and try to fit the various pieces together. It was like separating the most essential pieces of the jigsaw puzzle and then putting them together to create a picture of the total business environment.

Most importantly, writing the first part of the report helped me formulate the final forecast assumptions. At the beginning, I often needed a clearer idea of what was happening and what the assumptions would be. It was only after compiling and analyzing all the information that the assumptions were extracted.

Initially, the report should be distributed to the forecasting team. Once it is out there, people from other departments, including The Cheese, will want a copy. The report will be the best summary available of the company's business conditions. It will establish you as an expert on the business environment. It will also communicate to a broad group of people what the forecast is based on. This is important because:

The forecast is not wrong, but sometimes the assumptions are.

If you do not have time to write a full report or if your boss doesn't support the concept, it is still vital for you to document your assumptions every month, even for your own benefit and for discussion in the Sales & Operations Planning/Forecasting Meeting.

Benefits of Documenting and Communicating the Forecast Assumptions

Here are the reasons why you must document your forecast assumptions:

1. It is the logical basis and support for your forecast. The forecast and the assumptions are inseparable. Without providing supporting assumptions, your forecast is easily toppled and blown away.

2. The assumptions communicate to the forecast customers and other interested parties the state of the business and the factors impacting company revenue. They provide people with understanding and market

knowledge. They give people throughout the company a shared perspective.

3. It communicates and alerts people to significant changes to the prior forecast, explaining and supporting those moves.

4. It significantly reduces surprises for The Cheese, should reduce company conflict, and generates discussion of market factors. (Okay, maybe this one is wishful thinking.)

5. It can warn people about trends that threaten the business. Conversely, it can identify opportunities.

6. It provides the basis for analyzing your forecast accuracy and helps you refine your forecast process.

The forecast assumptions are vital to the forecasting process. Because of their impact on the numbers, they are almost as important as the numbers themselves. You will discuss the assumptions in the S&OP/Forecasting Meeting. When your forecast is challenged, you will discuss the forecast assumptions before you discuss the actual numbers. If you must change the forecast due to interference by The Cheese or other factors, you will only change the forecast after first changing the assumptions and documenting that change.

Having rational, logical, well-researched, and clearly communicated assumptions are vital for enabling you to stand your ground.

Reviewing The Forecast Assumptions

After measuring your forecast accuracy for the key month (Chapter 13), you will review your forecast assumptions (say your Forecast Assumptions report three months ago) tied to that forecast.

If your forecast was out of the accuracy standard, it is essential to know the reason. And almost always, that reason is a wrong assumption or even several wrong assumptions. This information will be documented in the forecast accuracy report (Chapter 13). A cursory review of the assumptions is still required if your forecast is within the standard. You may have gotten lucky and had multiple flawed assumptions that canceled each other out. You will want to know how good or bad your past assumptions were when developing the assumptions for the next forecast.

1. Therefore, managing the forecast assumptions is a continuous process as follows: The total market analysis produces the forecast assumptions.

2. The forecast assumptions are documented.

3. The forecast assumptions corresponding to your key forecast month are reviewed and evaluated monthly.

4. The forecast assumptions are revised, updated, and amended for the current month.

5. Repeat.

Relevant Story #2

Writing a Forecast Assumptions Report can establish you as an expert within the company and even throughout the industry. When I first started writing the report, the distribution was limited to the forecast team. However, people throughout the company began requesting a copy.

When the sales manager requested it be sent to the salespeople, I agreed but emphasized that it should not be given to customers because it contained proprietary information about our company. The salespeople benefitted greatly from the report because it made them look like industry experts. This led to deeper discussions with customers about the state of the industry and allowed them to report more market intelligence back to me. Talk about a win-win!

The salespeople asked for a way to report the information directly to their customers. So, I created a different "external" version of the report, including just the economic/industry information with no company proprietary stuff. We realized that the customer version would end up in the hands of our competition. We decided we didn't care and let it rip.

The customer report was a big hit. The salespeople made sure the report was sent to all their customers. It was one of those ancillary benefits that strengthened customer relationships. Since my name was on the report, it established me as an expert in the industry and helped me land the job at my final employer several years later.

Unfortunately, the customer report can come to an abrupt end. One of our largest customers, that "insane" one, again, became enraged (not exaggerating) that I had pointed out that the rapidly rising price of a key commodity *could* eventually lead to higher product prices and slower sales. This broad generalization led to the High Cheese at the customer screaming to the High Cheese at our company. Fortunately, my bosses defended me against the accusation of irresponsible behavior. But the decision was made to eliminate the customer report, which was a big disappointment to everyone except the insane Cheesehead.

Relevant Story #3

Do not disregard the value of experience ...

Sometimes, you may also want to document separately (not in your report) the assumptions of others that differ from your own. Note: This also applies to finding your expert contacts.

When I began managing the forecast process at my final employer, the long-retired company founder, Ed, was part of the forecasting team and attended our forecast meeting. Ed developed an extremely complex forecasting model in the 1980s, which continues to deliver outstanding market forecasts today.

The forecasting meetings consisted of me presenting all the market metrics, reporting all information from my contacts, analyzing the forecasting model mentioned above, stating the assumptions, and then discussing a proposed forecast before deciding on our final number.

Near the end of one of these meetings, I thought we had a general consensus on the final numbers when Ed said, "Your forecast is too high because of this factor." (I can't remember the exact details of this.)

We discussed this factor briefly, but I quickly dismissed it because my slew of charts, graphs, information, and assumptions indicated it was highly unlikely to occur.

Ed had not provided anything to support his statement, and I thought, *"Nice theory, old man, but you have to bring it stronger than that in my forecast meeting! I got charts, graphs, and analyses that say you are wrong."*

Four months later, I struggled to figure out why our forecast was too high. As you can tell, I take forecasting seriously and personally. Our customers paid us to provide accurate forecasts, and I always wanted to be more accurate than our competition. So, figuring out why a forecast is off is always part of the process.

It was not readily apparent why the actual numbers were significantly below the forecast. It took me digging deeper into the data and eliminating some possible factors. I finally identified the issue. I wrote it down on my notepad so I could report it in the upcoming forecast meeting. Then I realized I had heard that exact factor expressed by somebody months ago. I stared at my notes in disbelief.

Yes, that old guy had a tremendous "gut feel" based on his nearly 30 years of experience. I would never quickly dismiss his unsupported comments again, but I would document them. He wasn't always spot on, but he was never totally off-base.

Chapter 12
Getting To Your Forecast Numbers

As stated previously, this is not a traditional forecasting book. We will not delve into the numerous specific and technical ways to get the forecast numbers. There are dozens of books available to help you do that. In most of those, only 20-40 pages will be useful for your forecasting situation. I'm not a mathematician or a statistician, so you will have to look elsewhere to get that.

Here, we will examine the upper-level approaches to formulating a base forecast: Bottom-Up (BU), Top-down (TD), and Pure Statistical.

Bottom-Up Forecasting (BU)

This method involves getting your customer's monthly forecasts and then accumulating the data to generate your forecast. It works best when there are a small number of large customers who have a vested interest in providing you with accurate data. It doesn't work well if you have many small customers who do not get a big benefit from sending you forecasts.

If you are not currently doing any BU forecasting, it is a laborious process to set up. You must ask the customers to routinely send you their product forecasts. Their main concern is that their data remains private. By the nature of the request, they know you are asking their competition for the same data. You will need their commitment to providing the monthly forecasts and their trust that it will be kept proprietary. Your pitch to them is that providing the forecasts enables your company to offer better lead times and on-time deliveries.

Next, a process is needed to collect, record, and accumulate the data. Reminders must be sent when the customers forget or are delayed in sending in their data.

Setting up a BU forecasting system is usually worth the effort. It enables you to establish industry connections and will provide insight into the market's direction.

Another important reason for doing BU forecasting is that The Cheese expects you to do forecasting this way. They believe your job is easy. "Just contact all the customers. Get their forecasts. And bada boom – bada bing – we have a forecast." Some Cheese will disregard your entire forecast if you haven't done any BU forecasting. So, even if a BU analysis is only a minor factor in your forecast, I still recommend doing it.

Of course, BU forecasting has its flaws. You are depending on the forecasting ability of a group of other people (the forecasters at the customers) you have never met. You don't know their processes, ability, biases, or accuracy. Are these people quickly pencil-whipping the numbers as I did when completing my first forecast? Or are they master forecasters with excellent processes in place? If you rely on pure BU forecasting, it is not your forecast but a collection of others. In effect, you will have lost control of YOUR forecast, which is unacceptable.

Pure BU forecasting relies on the high forecasts to cancel out the low forecasts, and all their forecast biases cancel out, producing a great forecast, but that rarely happens. There is a substantial benefit, and no cost, for customers to over-forecast. There is a considerable risk for them to under-forecast. I pity the fool customer who gave you a forecast for 200 units for

July but actually needed 250, and your production constraints resulted in shipments of 210.

Therefore, raw BU forecasts tend to be considerably too high. Nonetheless, they help determine the market's direction and provide a key marker when finalizing your forecast. And, oh yeah, The Cheese likes them and doesn't understand why there are any issues at all with this method.

If you rely solely on Bottom-Up forecasting, you will want to record your customer's forecasts and measure the results monthly. This will help you adjust the forecast. You may be fortunate and find your customers tend to over-forecast 15% each month. Adjusting the BU number by that percentage could produce good results.

You may even identify a Master Forecaster among your customer group whose forecast is very accurate every month. You might base your BU forecast on their forecast. During stable economic and market share conditions, you could take their presumably accurate number and factor in that customer's average market share to get a market forecast. If their forecast is 4,200 units for May, and they average 20% of your sales, then the baseline BU forecast might be 4,200 divided by 20% or 21,000 units.

And if you find that Master Forecaster, get that person on your expert list. Find out what they know about the market and how they forecast so well!

Top-Down Forecasting (TD)

TD forecasting requires a reliable forecast or estimate of the total, upper-level demand for your products. Depending on your industry, that total demand number can be easy or difficult to obtain. At my final employer, we provided North American production forecasts for heavy-duty trucks, medium-duty trucks, and commercial trailers. The forecasts were by quarters, and the history was monthly data. Our clients paid for the information and used it in their TD forecasts.

For some industries, the government provides usable aggregate data. In other industries, trade associations offer data free or for purchase. It becomes more challenging when functional industry data is not available. You may have to purchase product registration data or other commercial databases to obtain total potential demand. Another option is to get numbers related to your industry and make assumptions to estimate your market.

Often, you will have to extract the portion of the market available for your products from the "total" market. These are assumptions you must make, and they can impact the accuracy of this segmented data, which in turn can affect the accuracy of your forecast.

You then multiply the adjusted demand number by your estimated market share to get your TD forecast. The estimated market share will be based on history, trend analysis, and information obtained from your research. So, if the total market/industry demand is 30,000 units and your market share is estimated at 20%, then your TD forecast is 30,000 times .20, or 6,000 units.

Statistical Software Generated Forecasts

Only a few industries are predictable enough to rely solely on statistical forecasts. If you are fortunate enough to work in those industries, you probably are not reading this book because you don't need a process, assumptions, etc. Even if you rely heavily on forecasting software, I recommend employing this book's concepts to improve your process, increase your forecast accuracy, and manage the corporate politics.

If you do use forecasting software as a primary or even a secondary method, never blame the software for an inaccurate forecast under any circumstances:

Cheese: "Your forecast was way off this month!"

You: "Yeah, the software didn't do a very good job this time."

Cheese: Thinking – *Maybe we could just get an accounting clerk to run the software and save some money ...*

Regardless, even the best statistical models or statistical software output need some adjustments. Our forecasting model at my final employer was highly complex, taking in over a million data points and being refined and updated over 40 years of use. It was the baseline for the forecast, and that number was then adjusted based on market conditions and factors outside the model.

High-powered forecasting software packages are essential when dealing with thousands of SKUs and a need to calculate component needs based on the data. However, that upper-level forecast is critical because those numbers will impact every descending level.

Forecasting software is also helpful for long-term forecasts. At some point, you will be asked to provide a 5-year or more extended forecast. The most efficient way to do this is to let the software develop the base forecast and then adjust for new products, accelerated market trends, etc. At one of my jobs, my monthly product family forecasts went out 12 months. I forecasted the first six months using the process described in this book and then let the software generate the last six months.

My five-year product family forecasts were primarily software generated beyond the current year, with minor adjustments.

Which Method Is Best?

The best forecasting method depends on your company, industry, and the data quality available to you. If your customers provide accurate forecasts in your key forecast month, you will rely heavily on BU. Even if they don't, you should collect the data from them and do a BU forecast. This is because, as stated previously, The Cheese expects it and believes this is the best forecast method. Even if you have conditioned The Cheese that BU forecasting is inaccurate for your company, you can have an issue with New Cheese if you are not doing this.

You will rely more on TD if you have accurate upper-level industry data. Of course, accuracy will suffer if the industry numbers are off. Your market intelligence data is more important here since you will adjust for market share shifts and special circumstances.

As stated, if your demand is highly cyclical and not significantly impacted by economic shifts, forecast software can offer consistently accurate forecasts. Even in predictable industries, it is better to use the output as the base forecast and then adjust it according to circumstances.

The Multi-Forecast Method

I recommend doing three forecasts and then deciding on the final one. Do the BU based on your customer forecasts and estimations of the rest of the market. You may not want to calculate the final number yet to avoid biasing the TD. Next, develop a TD utilizing your best data, market share estimates, and forecast assumptions. Finally, run a software forecast.

Now you have three forecasts, and you can triangulate, not average, the numbers. Theoretically, the best forecast number should be between your BU and TD. If the range between the two forecasts is narrow, it gives you confidence that your assumptions are valid. In that case, you can choose the method with the most historical accuracy or split the difference. Use the statistical forecast as a third check. If the software is usually 15% low, use this as a guide in checking the other two forecasts and triangulating the final forecast.

I understand that the multi-forecast method is laborious and can be considered overkill, but it prevents you from making big mistakes – and your next big mistake may be your last. It enables you to study every side of that forecast Rubic Cube and think through all aspects of the forecast. Sometimes, it reveals factors you may have missed.

If you forecast 15 product families, you will not have time to multi-forecast 15 times. Maybe you generate the three forecasts for the most critical product families and use formulas to break down the rest.

Measure the accuracy of the three methods monthly against the actuals and eventually determine which method works best. Under stable conditions, the statistically-generated forecast may even be a winner.

Finally, the multi-forecast method establishes you as a Master Forecaster and helps you gain credibility with The Cheese, which you will need when others criticize your forecast accuracy.

Cheese: "Well, did you ask our customers what they were going to buy?"

You: "Yes, I collect data from the top customers and do a bottom-up forecast every month."

Cheese: "But did you consider the industry and market factors?"

You: "Yes, I also estimated total industry demand and did a top-down forecast."

Cheese: "Did you do a statistical analysis of the market trends?"

You: "Yes, I run our numbers through a statistical software package every month."

After this discussion, you can proceed to defend and explain your forecast. (Chapter 15)

Running the numbers from various perspectives expands your perspective of the final forecast. This *total* perspective is essential when you must stand your ground.

Relevant Story

This has to do more with the incompetent people you encounter throughout your career than anything else, but it fits here best.

After attending an Institute of Business Forecasting and Planning conference, I wanted to incorporate forecasting software into my process based on recommendations there. I pitched the idea to my boss. He liked the idea but said, "I want you to work with the new VP of IT on this."

I looked at him incredulously. This was a minor purchase with no need for outside help.

"Yes, Andy is new and needs something to do. He just asked me if we had any projects for him, and now we do!"

I nodded, but my expression only changed a little. I left his office happy that the project was approved but knowing actually purchasing it was now much more complicated. When you work at a company long enough, you can smell trouble long before it arrives.

So, I met with Andy and explained what I needed. Two weeks later, I followed up, and he said he was still working on it. Two weeks after that, the response was the same. Gee, for someone who had nothing to do, it took a long time to progress on a simple project.

But then Andy announced that he was bringing in a potential vendor to give a presentation on his forecasting software. He didn't provide the company name, so I could not research the firm.

I arrived at the meeting and greeted the software dude. He was sporting an expensive suit as he set up his elaborate PowerPoint presentation.

As Andy and the guy exchanged small talk for an extended time, I wondered why the meeting hadn't begun. Suddenly, The Big Cheese entered the room, and the meeting commenced.

What the heck? I thought. This is a basic project, and now the Big Cheese is involved? So, I'm in this potentially important meeting with The Big Cheese, and my boss isn't here to provide me cover. I start to sweat.

This was the most bizarre and definitely the shortest meeting of my career. The software dude got all the way to slide #3 and was explaining the plethora of things his software could do, of which we needed none, when The Big Cheese barked out, "How much does this thing cost?"

The software guy doesn't blink and replies, "Our base package starts at $33,000." (Adjusted for today's dollars.)

The Big Cheese says, "That's way too much for us!" and zips out of the room.

I am stunned. The Big Cheese was correct in his assessment, but as usual, he displayed a horrendous lack of charm.

I was embarrassed and felt terrible for the software guy who had prepared the presentation, traveled to see us, and never even got to slide #4. There was an awkward silence after The Big Cheese left. I said goodbye and quickly scurried back to the security of my office.

After this disaster, Andy ignored my emails and gave cryptic answers in person. I researched forecasting software on my own and decided to go with the most basic one. We could always get a more robust program in the future if needed. I sent my recommendation to my boss and copied Andy.

Andy immediately responded (he didn't ignore this one!) that he "strongly disagreed" with my decision but offered no reasons or alternative suggestions. That's because Andy was a moron with a VP title – we've all known a few! Soon after he joined the company, they put a lovely new plant in his office, but it died in a few months. People joked that it couldn't live in all that stupidity. I said Andy was happy the plant died because he was now once again the most intelligent living thing in that office.

The software package worked fine for our basic forecasting needs. I used it to validate my monthly forecasts and for the 5-year forecasts. After five years, I upgraded to the next highest software package, which still cost a fraction of the price of that fancy package – oh, and Andy didn't make it to his second-year anniversary.

An Additional View

I wrote a white paper at my final employer (which I have included below) to help our customers use our data better and improve their forecasts. There may be some repetition here, but it provides practical examples. Even though it is industry-specific, think about how to apply these principles to your situation.

Top-Down or Bottom-Up Forecasting… Which Method Is Best for Your Company?

The debate on whether to use a Top-Down (TD) or Bottom-Up (BU) forecasting method has transpired for many years. There has been little consensus because the best approach will vary by individual company and product characteristics. While TD may work great for Widget Inc., it might be very ineffective for Gadget Corp. Most analyses of forecasting methods attempt to cover all markets and all products; this whitepaper will focus on manufacturing components that eventually go into producing commercial trucks and trailers. This analysis would also apply to industrial-type manufactured products and components goods in other industries.

The Methods:

Top-Down: In TD forecasting, you start with an upper-level forecast or estimate of the total market. You then make assumptions about various demand factors to determine the aggregate sales for the period. For example, Widget Inc.'s products are used in manufacturing heavy-duty trucks, and the forecast is for May. This company provides a forecast for the second quarter build (Factory Shipments) of 72,000 trucks. Step 1 – Calculate a build for the month. There are 64 working days in the quarter, so 72,000/64 = 1,125 per day. 21 days in May x 1,125 = 23,625 trucks in May. (This method typically works fine unless the market is changing rapidly.) Step 2 – Widget Inc.'s products are usually used on 74% of all newly manufactured trucks. The calculation is 23,625 x .74 = 17,482 or the "available market." Step 3 – Widget Inc.'s market share is averaging 36% and is expected to remain constant. The calculation is 17,482 x .36 = 6,293. Therefore, the aggregate forecast for May = 6,300 (rounded).

Production people will usually need more detail than this, so the aggregate forecast will need to be broken down into logical (by production standards) subgroups using history and market analysis. The optimal number of subgroups ranges from 6 to 15. It is difficult for the forecaster to analyze and break down more than 15 subgroups. Beyond that, formulas or other automation is needed, which can impact forecast accuracy. It is also usually difficult for production to handle subgroups that are too diverse, so 6 is a reasonable minimum.

The subgroup forecasts are then delivered to the demand or production planners for further delineation. Just because the forecast is done at the upper level does not mean the forecaster does not have to gather and analyze all relevant market information. This information will help the forecaster make assumptions supporting the forecast and be included in the Forecast Assumptions Report.

This market and company information will enable the forecaster to adjust the usage and market share assumptions used in the example above. Let's say Widget's product usage can vary between 70%-78% each month. In addition, its market share can range between 31%-41%. In May, very positive factors happened for Widget and product use and market share both hit the upper limits: 23,625 x .78 x .41 = 7,555 actual. If the forecaster failed to do a thorough market analysis and relied on just historical averages for the formulas, demand was under by 20%. The forecaster may know afterward why the forecast was inaccurate, but it is much better to do the work necessary to provide greater forecast accuracy.

Bottom Up: In BU forecasting, products are forecasted at a lower component level and then added together to calculate the subgroups and totals. The component forecasts are developed by one or more of the following methods:

1. Asking customers to send in forecasts at the desired components level every month and then adding all the customer data together.

2. Requiring salespeople to send in forecasts at the desired component level every month and then adding all the salesperson forecasts together.

3. Forecasting at the desired component level based on history and backlogs.

4. Using statistical software to forecast each component.

It should be noted that these methods can be employed without gathering any additional market information. The market information is already built into the customer and salesperson forecasts. The aggregate numbers may indicate what is happening with demand but may not reveal why it is happening.

Management is usually very interested in the "why" information, so gathering additional research is highly recommended. Methods 3 or 4 can also be employed, and marketing information can be used to adjust the numbers at the component or sub-group level.

Advantages and Disadvantages of the Methods

Advantages of Top-Down Forecasting:

1. It compels you to gather additional information and to understand the "big picture."

2. Understanding how the forecast was developed and presenting the results to management is more straightforward.

3. It is a cleaner, less complicated process, with fewer people involved and less data processing.

4. Making changes due to industry fluctuations or assumption adjustments is easier.

5. It is much better for developing long-term forecasts.

6. It has the potential for greater accuracy due to forecast errors getting canceled out at the higher levels.

Disadvantages of Top-Down Forecasting:

1. Requires accurate "top-level" industry data.

2. Can produce inaccurate forecasts if assumption percentage factors are subject to wide variations.

3. Can be accurate at the upper level and inaccurate at the lower levels.

4. Process is open to criticism by other departments (especially production) and others who prefer the bottom-up method.

5. Requires additional assumptions when segmenting the data regionally or by other criteria, which can affect accuracy. (For example: The Northwest Territory is 17% of the total.)

Advantages of Bottom-Up Forecasting:

1. Captures market variations that impact the forecast that might otherwise be missed.

2. Many managers and plant personnel prefer this process (just collect the information from the customers, and that's your forecast).

3. Better at providing more granular data at the component level.

4. Better forecasting of new products.

5. Better at providing segmented forecasts for factory purposes (30% of production should occur at the Buffalo plant) or regional distribution.

Disadvantages of Bottom-Up Forecasting:

1. It is a tedious and time-consuming process to collect and compile all the data.

2. Customer forecasts can be inaccurate (the forecast now relies on the customer's forecast process) or intentionally biased.

3. Salesperson forecasts can be inaccurate and biased (they may over-forecast and under-forecast on purpose).

4. Adjustments are needed to complete the data collected or when outside forecasts are late or absent.

5. A system is needed to process the data into meaningful information and to aggregate it into the higher levels.

6. Errors in the data collected or in the processed data are challenging to recognize, difficult to find, and sometimes difficult to fix.

When To Use Each Method

Use Top-Down Forecasting:

1. When there is good upper-level data available.

2. When there are many customers and/or sales are not highly concentrated among customers.

3. When doing long-term forecasts.

4. When there is data and history available to break the forecast into subgroups accurately.

5. When you have a process to collect industry/customer information that enables you to make valid assumptions.

Use Bottom-Up Forecasting:

1. When you have a small number of customers or sales concentration is high (greater than the 80/20 rule).

2. When forecasts provided by customers and/or salespeople can be trusted to be accurate.
3. When forecasting new products.
4. When systems are in place to collect and process the data promptly and accurately.
5. When there is a history of wide monthly demand variation.
6. When production requires (not just "would like to see") granular detail.

What Is the Best Type for Your Company?

Every company's forecast process is unique based on products, markets, distribution, and company culture. Almost all forecasting processes can be improved upon. When it comes to using TD vs. BU, there are companies on both sides of the spectrum. Some companies estimate (guess) the total market, estimate (guess) the industry factors, and then estimate (guess) their market share. Then, they can't understand why the forecast was so far off.

Other companies (including some industrial ones) build massive, elaborate systems to collect and enter vast amounts of customer forecast data into "monster" spreadsheets that churn out pages of results. The workers responsible for completing this task are frazzled, stressed, and usually clueless on what the final forecast really means. While thorough, this process doesn't ensure accuracy and is inefficient and expensive.

Therefore, a world-class forecasting process produces high accuracy while being efficient from a cost and time standpoint. The degree of forecast accuracy required, and its cost depends on the costs associated with inaccurate forecasts.

Select the method based on the factors above, then refine the process as needed. However, a process for collecting and analyzing market data and information is still required to make necessary adjustments to both TD and BU forecasts. These adjustments will be based on forecast assumptions.

The Best System Is an Integrated System

Many "world-class" forecasting companies use a "combination" approach, which involves doing both a TD and BU forecast and then resolving the differences into a final forecast. Statistical software may even be used to produce a third forecast data point. The advantages of a combination approach are:

1. It provides a built-in check system and can prevent major forecasting errors and miscalculations.

2. It will indicate which system, TD or BU, works best for your company.

3. It identifies unusual circumstances and factors that may have been missed or overlooked.

4. It identifies forecast assumptions that need to be reconsidered and modified.

When using this approach, it is advisable to produce the TD forecast first, especially if the BU data contains customer or salesperson biases. If the BU forecast is run first, these biases may influence the assumptions in the TD forecast.

When using a combination approach, the most important thing is to measure the accuracy of all three (TD, BU, and final) forecasts. Over time, the biases in the BU forecast and the issues with the assumptions in the TD forecast can be identified and quantified. Then there will be a basis and strategy for resolving and adjusting the final forecast, producing improved forecast accuracy.

****

This white paper was well received by our customers, who appreciated that we were willing to go beyond the numbers to help them forecast more accurately. It is critical to find the best method for you and your data, then refine and perfect it as much as possible.

CHAPTER 13
Measuring And Reporting Forecast Accuracy

Forecast accuracy is the acid test of all your efforts. The forecast is never wrong but varies in its degree of accuracy. This statement hinges on measuring that accuracy and determining its acceptability. Knowing your past performance and the reasons behind it allows you to stand your ground on current forecasts.

That's why measuring, analyzing, and reporting the accuracy of your forecasts is so essential to the process. The following is a basic example of the concept; modify it based on your situation, data, and statistical ability.

Product Family Results

You have designed your *initial* process to produce the most accurate forecast possible. You have identified the 6-15 product families to forecast. You have negotiated the long-term accuracy goal with Manufacturing. You have set your initial accuracy target and key-forecast month.

Now, once the monthly results are in, you measure your forecast accuracy against the target. Grab them and run your analysis as soon as the production/sales data is available. If your accuracy is out of standard, you must be the first to know and understand why before the relevant parties—or, in the worst case, The Cheese—point it out to you.

You will have spreadsheets set up where you plug in the monthly production/shipment numbers, which calculate the results for all months/periods of the forecast.

Here are the forecast accuracy results for Product Family B for April 2023. The January forecast was the key forecast, and it was 13.9% below the actual. If the forecast accuracy standard is +/-14%, this forecast just squeezed in the zone.

Product Family	B					
	1-Month	2-Month	**3-Month**	4-Month	5-Month	6-Month
Apr-23	+2.3%	-7.8%	**-13.9%**	-15.8%	-20.2%	-20.5%

Note: Set up your spreadsheet however you want since we all have our own style and complexity. I am only showing one row for discussion purposes. The actual sheet would contain all the rows from when you started measuring accuracy. So, you can compare how you did on the 3-month forecast in March and prior months. It is beneficial to look at the current row in the context of recent history. For example, if you have been under-forecasting a product family for the past three months, this indicates growth in demand that you are failing to pick up in your market/data analysis. You should factor that into the next forecast.

The most basic calculation of forecast accuracy is (Forecast Value – Actual Production/Shipments) divided by Forecast Value. In the example above, let's say the forecast for April 2023, in January, was 5,000 units, and the actual came in at 5,694 units. 5,000 – 5,694 / 5,000 = an under forecast of 13.9%. If your accuracy goal is +/- 14% at three months, you are just within the standard. You are just out of standard if the goal is +/- 15% at four months.

Even though you are measuring accuracy for six months of forecasts (or more if appropriate), you are only concerned with your key-forecast month because that one impacts Manufacturing or your internal customers (based on your discussions) the most. That is the standard you are measuring against. The 1-month and 2-month forecasts are typically irrelevant since Manufacturing can't react that quickly. Be advised that Purchasing may see the 5-month forecast as most important, but you should still use the key-forecast month dictated by Manufacturing as the standard.

Obviously, numerous more sophisticated statistical measurements can be used for measuring forecast accuracy. These are beyond the scope of this book. You will need to reference a traditional forecasting or statistics text. Use the measuring technique that enables you to improve your accuracy the most. Just win, Baby! I would caution against sharing the more sophisticated statistical measurements with people lacking mathematical acumen. It will confuse them and may be interpreted as being somewhat defensive when defending your forecast.

Top Level Results

A way to measure your total forecasting performance for the month is to compare the number of product families within the standard vs. the number outside. It is difficult to hit the standard for all product families you forecast in a given month. Generally, it's a good month if you are within the standard on 75% of all forecasted product families. If you only hit 50%, process improvements are needed. Sometimes, it is challenging to forecast the smaller product families, especially those with wide demand swings, but typically, these are less important and will have a higher materials safety stock.

Unfortunately, sometimes the upper-level performance doesn't matter much. If you are within standard on 11 of 12 product families but miss the most important one by 20%, your performance that month may not be highly-rated by Manufacturing.

Find the Gaps

The next step is determining why a product family was out of standard. You will refer back to forecast assumptions made three months ago as a starting point. Are there any faulty assumptions that impacted demand in that product family?

For example, the economy is now growing at only 1.5% instead of the 3% rate expected three months ago. This has restricted consumer discretionary spending, reducing demand for Product Family D.

Or, Customer C pulled up their orders by three weeks due to some unexpected maintenance issues at the plant. This boosted June's shipments but will lower July's.

Next, review the sales data per customer. What customers under/over performed and why? If you identify unusually high/low sales activity for critical customers you did not account for, you should contact their salesperson to determine why.

You may find out that a competitor is running a big promotion at Customer B or that Customer D has a new customer that is boosting sales for the product.

When asking the salesperson for post-forecast information, the question you never want to hear but often do, is: "Didn't you know about that?"

Inside of you, you are screaming: *No! No, I didn't know about it! And I didn't know because you didn't tell me. And it messed up my forecast! Arrrrrrrrrrrgggg!*

But you never say that out loud because you still need the salesperson "on the team". So instead, take a deep breath and say, "No, I didn't know. If you could inform me about stuff like this in the future, anything that impacts the forecast, I would greatly appreciate it because it helps the forecast to be accurate and the factory to run smoothly."

A worst-case scenario is when a salesperson calls you because your forecast has caused problems with a customer order/delivery, and their lack of communication may have been a contributing factor. If *you* made an error, admit it. If the assumptions were off, explain. If you didn't have all the needed information, the salesperson needs to be reminded about the concept of "team" forecasting and their role in the forecasting process.

And, of course, some things will happen outside of your knowledge and control that cause the forecast accuracy to be out of standard. It may be helpful to attempt to quantify the impact of these unusual events to explain the forecast performance or evaluate the base forecast's accuracy. Regardless of the unexpected events, a miss on the forecast accuracy is still recorded as a miss.

You must document the faulty assumptions and reasons why the forecast was out of the accuracy standard for all appropriate product families.

Because:

WHEN YOU "MISS" THE FORECAST, YOU NEED TO KNOW WHY – AND YOU NEED TO BE ABLE TO EXPLAIN IT.

If you are narrowly outside of the standard, there are usually no significant issues. But when you whiff big, you can get a visit from a displeased internal customer. You may get that angry call from the salesperson with a legitimate gripe. You may have to provide an answer to your boss who has received a complaint about the forecast. And in the worst-case scenario, you may be confronted by a displeased Cheese. We will cover how these conversations should go in Chapter 15. But you can only defend the forecast if you have the needed weapons. The post-forecast review provides the armament. The other reason for the post-forecast review and evaluation is to provide information for the Forecast Accuracy Report.

Forecast Accuracy Report

Of everything recommended in this book, this is the most difficult thing for a forecaster to do. It is counterintuitive, carries some risk, and is controversial. It's not that compiling a Forecast Accuracy Report is difficult, but distributing the report takes guts. It's not for the wimpy forecaster – it's for the Master Forecaster.

But don't worry; the distribution of this report is the culmination of this whole revamping/revising of the forecasting process. It is accomplished after you have become a 4-C (Competent, Credible, Confident, Controlled) Master Forecaster. You will not distribute this report until you are confident in your process and your performance.

Report Contents

The report will contain charts showing your forecast accuracy against the standard for each product family for the past six months. It will also have a summary chart showing the forecast accuracy of all product families for the prior month. For each product family, write a few bullet points. For the product families that are outside of the accuracy standard, list the assumptions or factors that put the accuracy out of the accuracy standard. There is no need to go into excessive detail. Keep it as fact-based as possible. These are the reasons the forecast was off – not excuses!

For the product families within the accuracy standard, bullet points highlight the correct assumptions and whether these assumptions are expected to continue. It is important to remind people when a forecast is highly accurate because you will be told repeatedly when it is not. When you have an excellent forecasting month, celebrate it – it will help to offset the feelings of a bad month.

A good length of the report is 5-6 pages, including the accuracy charts. You want the report to be several pages because you want most people to skim through it and not study it in fine detail. The fewer questions, the better. Only a few people, usually the production planners, will read the entire report every month.

Here is an example of a basic tracking chart that would be included in the report. It tracks the forecasting accuracy of Product Family C over a 12-month period. The forecast accuracy standard is +/- 15%. The months outside the standard are denoted by the checkered pattern (they are shown black-and-white here due to book printing restrictions). Of course, on color charts the bars within standard are green, and those outside are red. I used blue bars when the actual was equal to the standard.

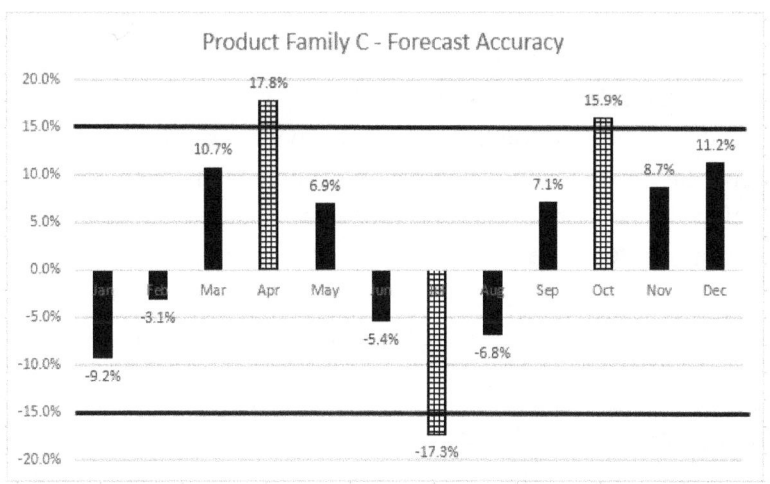

Purpose of the Report

1. Total transparency – No surprises

When the forecast is out of the accuracy standard, you will be the person reporting this, along with the reasons why. This presents an image of a competent forecaster in control of the situation.

Being transparent allows you to control the message and greatly reduces getting blindsided by attacks. There should be no arguments about how "bad" the forecast was since the results are explicitly reported for all to see. This should stop that weird jerk (every company has one) from blasting, "Wow, that October forecast really sucked!" when maybe it was actually within the accuracy. It should neutralize the "forecast was wrong" charge and lead to a discussion of how the assumptions impacted the forecast accuracy that month.

2. It is helpful information for Manufacturing.

Manufacturing must monitor your forecasting accuracy closely when developing production plans. Remember: Your job is to deliver the most accurate forecast possible. Their job is to develop the best production plan possible based on your forecast. Production planners need to know your forecasting performance, including trends and biases, to do their job effectively. This information can also be helpful to Purchasing.

If you have been under-forecasting Product Family F for the last six months (even if five families are within standard), the Production Planner may decide to bump up the production plan to account for your past performance.

Of course, you should also adjust for these same tendencies when making your next forecast. This is always part of your monthly data review process.

3. It communicates the results to the forecasting team.

If you have incorporated a forecasting team approach, it lets the team know how it is doing. The forecasting accuracy results can spark valuable conversations among the team before and during the Sales & Operations Planning/Forecasting Meeting. It is also an excellent opportunity to showcase some of the team member's contributions, thus building goodwill and strengthening the comradery and overall process.

4. It provides the first line of defense against personal attacks.

You have established a forecast accuracy standard, developed a system to measure against that standard, communicated the assumptions and the actual outcomes, and provided relevant information so that anyone who has a big problem with your performance can talk directly with you instead of behind your back.

Essentially, it moves the target from your back to your front. As a forecaster, you can never remove that target - it comes with the job. But wouldn't you prefer to be physically attacked from the front when you have some chance to defend yourself versus the back when you could be dead before you know it? You will have the weapons to defend yourself against a frontal assault because corporate backstabbing can be fatal to your career. Having this information already out there helps you stand your ground. As stated previously, forecasting is a perilous profession. Attacks can come from any direction at any time. Having your results out there, even in a weak month, is a prime defensive move.

In addition, now your boss knows your monthly process, so they should be able to give intelligent answers to The Cheese if needed. It should also make your performance appraisal smoother since you have objective data to discuss.

Publishing results can be difficult when you are in a *forecasting slump*. Forecasting is difficult, and sometimes, you can make logical assumptions, have valid analyses, etc., but miss your forecast accuracy goals. When the market is in transition, this can happen for several months until you are able to regain an adequate understanding of the environment. Highlighting your failures is not enjoyable, but it does gain you credibility. Credibility helps you stand your ground.

5. It improves your visibility.

Remember that one of the problems of being a forecaster is that people expect perfection and that you only become visible when you screw up? Publishing good results allows you to become visible and toot your own horn when applicable. It can balance out some of the bad with the good. Also, it highlights that you have an accuracy "standard". This means you don't have to be perfect, so it lowers expectations, which is a good thing.

6. **It makes you a better forecaster.**

How important is your forecast to you, knowing you will publish the results at a later date? This is the best motivation for producing an accurate forecast. Sure, you want to provide Manufacturing with good numbers, but the Forecast Accuracy Report with your name on it makes it personal. I would feel happy or sad the day I calculated my monthly results. It felt great to send a Forecast Accuracy Report with 12 green bars. I was annoyed when there were six red ones.

Before You Take the Plunge

1. **Make sure your forecasting process is improving your accuracy and achieving good results.**

You will measure your accuracy from the beginning of your process revamp/revision. During that time, this is your personal yardstick to measure progress and biases. You may share the results with your primary contact in Manufacturing during discussions, but you gain nothing from releasing a report with excessive red ink. It may take up to a year before you are ready to release a Forecast Accuracy Report.

2. **Discuss it with your boss.**

Before the initial release, you will want to discuss this report with your boss. You will need to explain the reasons for and benefits of this report. They may not permit you to release it. In that case, still compile the information but only use it in your forecast/data review.

Distribution

The Forecast Accuracy Report should be distributed to all members of the forecasting team and your boss. Initially, it is unnecessary to send it to any Cheese or other departments. Of course, others, including The Cheese, may request a copy and be added to the distribution list. Be aware that people may redistribute the report without your knowledge.

Questions

As the Forecast Accuracy Report is redistributed, you will get questions about it and your process. Eventually, a Cheese will appear with the report in hand, and some questions. This is a good thing. It enables you to explain your process and your commitment to forecasting accuracy.

Here are common questions and your responses.

1. **What do the red bars mean? How do you know the forecast is off?**

I talked with Bob Radcor in Materials, and he said they needed a 3-month window to react to demand changes, so we measure against the 3-month forecast since it is the most important.

I set my initial standard of = +/- 15%, and I have been able to tighten it to +/- 13% over time. The green bars mean I was within the accuracy standard. The red bars indicate that the forecast was outside the standard. The commentary explains why the forecast for the product families was either in or out of standard.

2. **What problems are caused when the forecast is out of standard – indicated by the by the red bars?**

Production Planning must respond to the variance when the forecast is out of standard. Small misses usually are easily handled. You would have to talk with them for more insight.

3. **What are you doing to improve your accuracy?**

Every month in the Forecast Accuracy Report, I identify which forecast assumptions were incorrect and how they impacted the forecast. This allows me to identify the areas I must concentrate on for future forecasts.

4. **Why are you doing this report?**

The report provides valuable information to Manufacturing about the accuracy of the forecasts. This gives them better insight into how they respond to future forecasts. It also helps them evaluate their performance in the past month. In addition, it communicates to the forecast team how we are doing.

Building Credibility and Competency

After hearing your answers, you hope The Cheese, or anyone else, comes away with the following impression:

This person has firm control of their job and is dedicated to improving their performance. They are competent and have credibility. (Or something to that effect).

You create "good vibes" that offer protection when things go amiss, which, by the nature of the job, they will. If your bosses know you have a good process in place, they will assume something unanticipated has occurred and not immediately blame you (even if you mucked it up!) when the forecast is off.

Risk Vs. Reward

Yes, there is a risk involved in publishing the report. However, I cannot remember any significant negative incident of publishing the report at my previous employer. This was in a stressful, highly political corporate environment and with four different bosses.

To repeat: Nothing motivates you to produce accurate forecasts more than knowing that these forecasts' accuracy will be self-reported publicly within the company. I knew that day when I calculated my forecasting results from the previous month, it would be emotional. I was either going to be happy or distressed. Both of these emotions are strong motivators.

You also gain tremendous credibility by being bold enough to publish your accuracy results monthly. Much of it is to create a positive image: Would someone doing a lousy job highlight their results?

So, to be a Master Forecaster, publish your accuracy statistics monthly. And be prepared then, to stand your ground.

For The Business Predictors

Develop some personal accuracy standards you want to achieve, and measure your actual results against it. It will motivate you and make you a better predictor by analyzing your failures and successes. I do not recommend issuing a report on your failures. Often The Cheese is looking for a scapegoat for their own failures, so don't volunteer to take the hit. If you have had a good year, bring your scorecard to your performance review in case things go sideways.

Relevant Story

Complete transparency has its limits ...

At a former employer, the Director of Quality asked me if he could include my forecast accuracy statistics in his monthly corporate presentation, highlighting all the critical metrics.

(This is a good example of how the Forecast Accuracy Report circulates. Although he was not on the distribution list, he was well aware of the report).

I declined the request because the data served no real purpose in the presentation but to provide another colorful chart to present to the staff. And I did not want The Cheese to ask questions about this without me being there to answer them. If I had agreed, I would have wanted to attend these meetings to quickly extinguish any potential blaze. I had much better things to do with my time.

Chapter 14
Refining Your Process – Tightening The Standard

You now have a forecasting process:
1. Collect market information and analyze internal data.
2. Develop assumptions based on your analysis of the available information and data.
3. Develop your forecast based on your preferred, most effective method or a combination.
4. Deliver your forecast and the supporting assumptions to your internal customers.
5. Measure the accuracy of your forecast against the actual numbers. (You will measure after refining your initial process and then distribute the results later, when you are ready).

This provides the basis for improving your forecasts. Despite all this hard work, your process still needs improvement. It has holes and gaps which need to be plugged. As you tweak and tighten your process, your forecast accuracy increases. The challenge is to keep refining the process until you achieve your accuracy target and beyond.

The Hockey Goalie Analogy

Imagine you are a hockey goalie. The opposing team called "Uncertainty" is firing pucks at you that can *ruin* your forecast. Your enhanced forecasting process enables you to make a stick save on one, a pad save on another, and a glove save on a tough one. But you left the five-hole (between your legs) open, and one puck slams against the back of the net. GOALLLLLLLLLLL! And suddenly, the forecast is outside the accuracy standard, and you have lost the *game* that month. Improving your forecasting accuracy hinges on finding the gaps in your process and sealing them. In other words, continually improving your forecasting process.

Establishing a Baseline

For example, your initial goal is to be within +/- 15% on the three-month forecast for your 10 product families. In the past, you averaged five out of ten product families within the standard for the month. Not an excellent performance, but that's why you developed your process.

The November numbers arrive, and after plugging them into your performance spreadsheet, you hit 7 out of 10 within the standard – making it an above-average month. However, Product Family G was 20% under the forecast. One puck, or several, made it past the goalie.

Next, you dive into the data to try to determine why. You compare the variance against the assumptions relevant to the product family, then ask the following questions:

1. What did I assume was going to happen on this factor?
2. Why didn't it happen in this way?

3. Why did I make the wrong assumption?
4. What didn't I know then?
5. Why didn't I know it?
6. What impact (estimated) did this factor have on demand?
7. In summary: What did I miss, and why did I miss it?

Remember: The forecast is never wrong, only inaccurate. Conversely, assumptions are often wrong. And in assumption-based forecasting, flawed assumptions produce inaccurate forecasts.

Of course, these questions can lead to more questions:

1. What essential data did I have that I deemed irrelevant?
2. What new economic/industry indicators do I need to track?
3. What data analyses do I need to add to my review?
4. What information/data sources would have helped me plug this gap?
5. What forecasting biases negatively impacted my forecasts?

Whenever you miss a forecast, you try to modify your process to plug the holes. Of course, sometimes events can't be predicted. Large order cancellations, warehouse fires, and freight disruptions will all mess with the forecast but are outliers when doing your analysis. So, let's focus on what we can control.

This is part of the continuous improvement system: Plan – Do – Check – Act. Make analysis/assumptions (Plan) – Develop forecast (Do) – Calculate accuracy (Check) – Tighten process (Act).

Another way to look at it is to identify the problem (gap). Identify possible solutions (plugs). Make changes (action). Measure the results (forecast accuracy) and repeat the process.

Checking the Information/Data Stream

Identify the Holes

What information/data did you lack that would have helped you accurately forecast the out-of-standard product family? What did you miss, and was that information/data available to you?

First, analyze the sales data. The best place to start is sales by customer. Was the variance caused by a few customers or across the board? Was the variance isolated to one or two sales regions? Then, the sales by specific products within the product family will be examined to narrow the search.

Once you identify the gaps, you can communicate with the relevant salespeople to gain clues about changes in customer behavior or product preference, which sunk your forecast. You hope you don't get the "Didn't you know that?" response discussed earlier. You also don't want the "I don't know" answer. However, the salesperson may not be fully aware of the situation since the data has just come out. They will appreciate a heads up and get back to you with important information after investigating.

Good salespeople will already know the answer to your question. Your main concern is if this occurrence was a blip or a trend. Near the end of the conversation, express that the reason for the call is that this event caused the forecast to be too low/high and ask for their help knowing about these events in advance. Typically, the salesperson will appreciate your effort in doing your job well. The Forecast Accuracy Report will include information gleaned from these conversations to explain the missed target.

In addition, were there any economic indicators that reacted differently than expected? Was GDP significantly higher or lower than expected? Were there issues with the key indicators you track that impacted demand? Did the government revise any relevant data? Did one of the "minor" indicators unexpectedly and significantly impact demand? The relevant information from your review is detailed in the Forecast Accuracy Report.

The key is to locate the holes in your information/data process. That's the easy part; the challenge is filling them.

Plugging the Holes

What relevant information/data didn't you have, and how can you get it in the future? You want to modify your data analysis/collection to improve your forecast accuracy.

For example, if demand at Company R swings widely, establish a reliable contact there and check in, at least by e-mail monthly. If sales at Company D are stable, you may only need to check in once or twice a year.

You may need to investigate new factors impacting your forecast more thoroughly. This could require finding a new expert contact in that area. For example, a rapid rise in steel prices is crimping demand. A steel wholesaler can provide insight into future pricing.

Are there economic indicators that have become more relevant? In the previous example, you may have to start tracking steel prices in the short term if they affect the forecast. At my final job, the relevant economic indicators were constantly changing, and the challenge was to identify the fresh ones and drop the stale ones quickly.

The critical thing here is that the relevant data/information changes as the market changes. You must modify your process to keep up with it and improve the forecast accuracy.

Identifying Your Forecast Biases

Everyone has biases; it's part of the human condition. And everyone has forecast biases. If you ask most forecasters if they have biases, they will deny it because they are invisible to them, similar to personal biases. But as a forecaster, you have biases. You may not know what they are, but I guarantee you, the users of your forecast do. Economists often miss their forecasts due to personal/political biases.

The good news is that your biases become evident by measuring your forecast accuracy against the standard. If you have under-forecasted a product family for six straight months, there is some bias causing you to under-forecast. Review your assumptions to see what "negative" assumptions are having too much influence.

The bias is more prevalent if you are consistently over or under-forecasting all product families. These situations often occur during pronounced economic cycles (think economic slowdown or robust recovery).

Once you identify your biases, you can adjust your next forecast to account for them. Suppose you had under-forecast a product family for several months. Then, you might consider raising your calculated forecast for the next month by 5%. So, one of your assumptions (not publicly stated) is that you have been under-forecasting the product family. You don't want to do this often, but it sometimes works and can break you out of one of those frustrating forecast slumps.

Planned Biases?

Shouldn't you hedge your forecast to the high side to account for safety stock, or shouldn't I be conservative to keep inventory low? The answer is "No."

Remember: Your job is to deliver the most accurate forecast possible. Their job (Manufacturing/Materials) is to take that forecast and develop the best production plan possible. Everyone is happy when both functions do their jobs well, especially the accountants.

You may encounter company political biases. For example, a Cheese may tell you not to over-forecast because that caused problems at his last company. If this happens, you could intentionally be conservative with your forecasts but change your accuracy range to +5% to -20%, for example. Just make sure Manufacturing is aware of the change in strategy so they can adjust their production plans accordingly.

Financial forecasts always have inherent biases. The corporate Cheese, the divisional Cheese, and the financial Cheese all hate financial forecasts that are too high because then it appears that the organization failed to achieve a realistic goal. Therefore, keep financial forecasts conservative – BUT ALWAYS INCLUDE FORECAST ASSUMPTIONS. In this case, you want final buy-in from a Cheese on these assumptions. If things go bad, you need as much cover from the fallout as possible.

Relevant Story #1

Reducing Forecasting Biases ...

When I began managing the forecasting process for commercial vehicles at my final employer, my company had a reputation for producing conservative forecasts. While generally more accurate than the competition, the company rarely over-forecasted. This is what I label "Price Is Right" forecasting—do a good forecast but don't go over (which is appropriate for financial forecasts, as mentioned previously).

Conversely, our competition was regarded as aggressive forecasters, usually overshooting the mark, sometimes significantly. So, in the industry, my company was regarded as more accurate when demand was receding, and the competition shined when the market rose. Companies subscribing to both forecasts told me that averaging the two produced great forecasts. (There's a simple process for you).

Now, I don't believe in conservative forecasting – I believe in accurate forecasting. However, my company utilized team forecasting, and I was the team's newest member.

I knew I needed to change the corporate culture, but corporate culture changes happen very slowly. To repeat: CORPORATE CULTURE CHANGES VERY SLOWLY. (A side note: You may have to change your forecast process incrementally and very quietly because of this).

In our monthly forecast meetings, I almost always lobbied for a higher forecast and achieved a few minor victories. The breakthrough came when I had argued for a higher yearly forecast number for three months in a row, and we finally settled on that exact number in the forecast meeting four months later. Of course, I jokingly pointed that out to the team, but the message sunk in: If we are less conservative on our forecast, we can be more accurate.

Going forward, our forecasts did become less conservative and more accurate. Our customers didn't notice the change until that fateful month when our forecast was higher, Egad! - than our competition's. The day after the forecast was released, my phone rang all day as I explained and defended our forecast to numerous customers. We were only 3% higher than the competition, but it was a big deal, at least for that month. From then on, we were no longer considered the most conservative – just the most accurate! Note: There are always some unexpected challenges when attempting to change corporate culture.

A few years later, the change in forecasting philosophy really paid off. The industry was recovering from a downturn (where, historically, our competition had an advantage). The competition and every other industry analyst were predicting a slow, gradual rebound. However, our analysis showed a much sharper, faster recovery. We boldly put out a forecast almost 20% higher than the general consensus – and we nailed it!

Reviewing Your Forecasting Methods

Suppose you run the forecast three ways: Top-Down, Bottom-Up, and statistical. In that case, you should measure the accuracy of all three forecasts to determine which method currently yields the best result.

This implies that the accuracy of the three methods is fluid. Sometimes, the assumptions for the Top-Down forecast are inaccurate because the demand drivers are shifting, throwing off the industry totals. Other times, the customers feed you the wrong numbers, sidetracking the Bottom-Up forecast. The statistical forecast can be way off or spot on, depending on market trends.

Therefore, the recent accuracy of the three methods determines what weight you give them on the next forecast. Suppose the Top-Down forecast is always higher than the actual, and the Bottom-Up is consistently lower. In that case, I have no objection to averaging the two forecasts (I occasionally did this). But, if this is your prevalent method, keep it from your boss or The Cheese because it sounds too simple, and it discounts the hard work you do on the Top-Down and Bottom-Up forecasts. If asked, your answer is always the same: "I run the forecast three ways to ensure I don't miss anything and then triangulate the numbers to get the final forecast."

Back To the Goalie Thing

There are holes in your forecasting process. Initially, you can't see them and don't know they exist. But uncertainty fires pucks at us every month that slide through the gaps and cause us to miss our forecast accuracy targets.

Therefore, every month, attempt to fill as many gaps as possible. Over time, you can block more shots, know more, and increase the accuracy of the forecasts.

Moving the Goalposts

Usually, the term "moving the goalposts" has negative connotations, but in this case, it means something much different. As you make advances in the forecasting process, your forecast accuracy improves. At some point, you should be hitting your accuracy target with a high degree of regularity. When that happens, it's time to tighten the standard. In other words, if you make 95% of your field goals with an 18.5-foot crossbar, let's move the goalposts to 17 feet.

When and Where?

When you change your accuracy target and by how much is your decision. They are your goalposts. You set the initial standard, and you can set the new standard. I was doing well when 9 of my 12 product families were within the standard for the month. I TIGHTENED MY ACCURACY STANDARD when I was consistently hitting the standard on 10 or 11 product families out of 12. Forecasting the more critical product families is essential. The smaller product families, or less important ones with wider demand swings, are not as crucial. In these situations, you may want to widen the accuracy standards of these product families to make your total performance appear better. Remember, these are your goalposts; you can plant them where they make sense.

For example, say you began with a 17% accuracy target for your 3-month forecast for 10 product families. Initially, you averaged around 6 product families a month within the standard. You start to tighten the gaps in the forecasting process. Now, you are averaging over 9 product families within the accuracy standard a year later, often going 10 for 10. It is time to tighten the standard.

How much should the standard be tightened in this example? A suitable method for determining this is to take the average of the forecasting variance for the last six months. Let's say the general average variance is 15.5%. Moving the target to 15% is reasonable. If it is 16.2%, it is safer to go to 16%, hoping to move it to 15% in three months. But what if the average is 13%? Well, either you are proficient at improving your process, or the market demand has stabilized. I would set the standard at 15% until, I was sure. You do not want to have to loosen your standard. It is difficult to explain to the forecasting team, your boss, and especially The Cheese how you are implementing continuous improvement but moving backward as a result. Remember that your forecast accuracy will suffer when market dynamics fundamentally change.

This brings us back to the Plan – Do – Check – Act of continuous improvement. You keep improving your process and measuring the results. As the accuracy improves, you tighten the standard until you hit the goal and then keep going, if possible. If you can hit the negotiated accuracy standard and then keep improving, you are a Master Forecaster. And when you are moving forward, it is much easier to stand your ground.

Relevant Story #2

Your internal customers like continuous improvement ...

The Materials Manager gave me a forecast goal of being within +/- 15% on the 3-month forecast. Since I was starting from scratch with no forecast history, that was my goal. I began building my process and measuring my results. Nine months after revamping the process, when the results showed noticeable improvement, my forecast accuracy report was published.

After two years, I consistently hit the target, especially in the key product families. I tightened the standard to 14%, then 13%, and finally to 12%. Even with process improvements, I could not move the goalposts any closer. You cannot achieve perfection. You cannot defeat uncertainty, no matter what The Cheese believes. My internal customers had initially requested a +/-15% range, and I was giving +/-12% most of the time. So, they were delighted with my forecasts, which is the ultimate goal.

Varying the Standard for Product Families

Fortunately, I could keep the same accuracy standard for all 12 of my product families. My products were highly homogeneous. The economic and industry factors driving demand were essentially the same. Even the parts going into the products varied little. Materials-wise, I could miss high by 5% on one family and low 5% on another and cancel out the difference on many components.

But there is no reason you cannot vary the forecast accuracy standard for each product family. It makes sense when the demand swings and variability differ. While it does make things more complicated, you can easily plug the new forecast targets into your measurement spreadsheets. You should note the accuracy standard on your accuracy tracking graph, and designate the target on all relevant spreadsheets and charts to help you remember. It might help to break your product families into subgroups, with Group 1 at an accuracy standard of +/- 15% and Group 2 at +/-13%, for example.

Communicating the Change in the Accuracy Standard

You do not need to make a big announcement regarding the change in the standard. It should be noted in the appropriate

Forecast Accuracy Report and reported in the forecast meeting. You can explain the change in the forecast meeting, which allows you to remind people how the target was established and what the goals are.

For The Business Predictors

You have a process in place designed to ensure you evaluate all relevant factors and improve your chances of making a correct recommendation, call, prediction, or analysis. However, there are gaps in your process. Because analyses are not usually standardized, there are more gaps than when doing monthly forecasts.

When you make an incorrect call, review all your assumptions. This is why it is critical to document the assumptions on which your recommendations are based. The initial analysis may have been conducted two years ago, and no one remembers what the business conditions were at that time. So, review the assumptions to determine which faulty ones sunk your recommendations. On long-term projects, new factors, which were unknown at the time of the analysis, could have had a negative impact.

It is vital that you know why your recommendation was incorrect. You want to fill in those gaps for your subsequent analysis. For example, let's say government regulations caused your whiff. In your next analysis, you will check to see what regulations have been discussed or suggested.

Secondly, you need to be prepared for The Cheese bursting into your office unexpectedly and demanding to know how you could be so stupid as to recommend this obvious disaster. You will need to calmly discuss the assumptions made at the time and how things did not go as expected. The Cheese has a convenient way of forgetting the past when it is unflattering to them. Maybe the assumptions listed in the report were theirs to begin with.

Plus, beware of the obvious-now syndrome – of course, everything is so obvious now when you know so much more – but they may have been highly improbable last year when the analysis was completed.

You may even consider distributing a "post-analysis" report detailing why the recommendation didn't go as planned. In rational companies, this will show how you are on top of things and a valued asset to the company, even if you failed on this one. However, in toxic ones, this just gives people someone to blame. Remember, just because you did the analysis, you are not responsible for executing the strategy and the ultimate results.

CHAPTER 15
Defending Your Forecast

If forecasting is "an art and a science," then defending the forecast is totally art, but supported by the science. The "art" here is to be able to stand your ground without generating heated arguments and long-term political damage.

The challenge is to effectively and vigorously defend your forecast – without being defensive. Yes, you read that correctly. For emphasis:

YOU NEED TO BE ABLE TO DEFEND YOUR FORECAST WITHOUT APPEARING DEFENSIVE

How do you do that? Well, that's why this chapter is here. For experienced forecasters, it may be the most valuable information in the book.

Limit the Arguments

As the old forecast meetings at a previous job displayed, forecast discussions can get emotional and heated. They can quickly spin out of control, turning into shouting matches. These are no-win situations for the forecaster. Even if you "win" the argument, you can end up losing big time. Lose enough times when defending your forecast, and you will lose your job. You need to be able to stand your ground without making political enemies.

It's similar to having an angry argument with your spouse. Have you ever won an argument but suffered through the long-term effects of the hurt feelings afterward? And you gain nothing by arguing with your boss or, especially, The Cheese.

They will conveniently forget if *you* are proven correct, but those negative impressions of you may last forever. Oh, but they will remember if *they* were correct, which means a trifecta loss for you:

1. You were wrong.
2. You didn't acknowledge/realize you were wrong.
3. You argued too aggressively. (Even if The Cheese was being rude and obnoxious).

Unfortunately, your coworkers, and especially The Cheese, may start the conversation by attacking and disparaging the forecast. Refrain from getting suckered into reacting back in kind. Take a deep breath and respond rather than react. If you display the 4 C's - Competence, Credibility, Confidence, and Control, you can respond to most challenges to your forecast professionally and productively and stand your ground.

Defending Your Numbers

Your forecast and the assumptions report have been distributed. Now, you may receive an email, call, or personal visit from someone who believes your forecast is too high or too low and should be changed.

This is usually someone outside the forecasting team since they would have had the opportunity to raise objections during the S&OP/Forecasting Meeting. Now, it still could be someone from the forecasting team who feels their previous objections were disregarded. Or maybe this person left the meeting and broadcasted their objections to anyone who would listen, including The Cheese. The worst-case scenario is to have The Cheese, maybe even The Big Cheese, appear out of nowhere clutching your forecast with a scowl on their face.

No matter who they are, you are likely to get the following volley:

"I think this forecast is much too low."

Or,

"I don't see how we are going to get to these very optimistic numbers in October."

In response, you want to refrain from discussing any numbers at this point.

Now, what do you mean you don't want to discuss numbers? This person, perhaps a Big Cheese, is standing at my desk to challenge my forecast. He wants to discuss numbers.

You want to avoid discussing numbers up front because it can quickly become an argument about the numbers. Then a pissing match about the numbers, and depending on the status of the challenger, your forecast may get pissed on.

Therefore, you want to discuss the forecast assumptions before you discuss the actual numbers. The assumptions are critically important to your forecast and in defending it. That's why it is essential to have an effective information/data collection process to develop solid assumptions. This is why you discussed the assumptions during the S&OP/Forecasting Meeting and issued a Forecasting Assumptions Report. The whole process fits together and provides the opportunity to discuss and defend your forecast calmly and professionally.

If the forecast challenge is delivered by email, you can respond with:

Have you read the Forecasting Assumptions Report? Attached is a copy. Please let me know which assumptions, or any new ones, you would like to discuss.

The response by phone would be similar. If the person has not seen the report, offer to send it to them and discuss it later.

In person, your first question remains: "Have you read the Forecasting Assumptions Report?" You must ask this question without any disrespectful attitude or smugness, especially if you are communicating with The Cheese, who hates it when the underlings cop an attitude.

Your aim is to start a productive conversation about the forecast by asking this question. You may wind up changing the forecast if the discussion provides new, relevant information.

You ALWAYS want to discuss the forecast assumptions, not the numbers. If the person has not read the forecast assumptions but wants to have a discussion (like when The Cheese is standing at your desk), get your report and read the assumptions to them. Ideally, if they disagree with the forecast number, they disagree with some assumptions.

This should lead to a discussion, not an argument, about the assumptions. This discussion is vital because the person can add insight, opinion, information, and knowledge you didn't have when you completed your forecast. You value these discussions because you can learn new information from them. This interaction can have positive aspects as you gain insight and establish credibility, even if the debate is contentious. Show them you are in control of the forecasting process and can discuss your forecast without being defensive.

At my final job, I told our customers that if we published a forecast they disagreed with, I didn't want them thinking we were terrible forecasters or complaining about the numbers to their boss. I wanted them to call me immediately and discuss it. Yes, I wanted people to challenge my forecast! Not all of those calls were enjoyable, but I learned something from every discussion and became more informed afterward. I also gained tremendous credibility with the users of my forecast. I would tell them, "The only thing worse than putting out a bad forecast is not knowing it is a bad forecast."

Sometimes, the person challenging your forecast agrees with your assumptions but has other assumptions and information that they base their opinion on. If these factors are correct, you may have to concede that the forecast may be off. Often, The Cheese, due to their network and access to details, may have information that is news to you. Which leads to the dreaded question – delivered with attitude:

"Didn't you know about this?"

At that moment, you feel just like that hockey goalie when he hears that puck snap against the back of the net and the horn sounds. It stings worse when the question comes from The Cheese because it implies you should have known, and of course, it's your job to know.

The best you can do at that point is:

"*No, I'm sorry, I didn't know that. Let me check with Chad to get all the details.*"

Yes, you are deflecting some of the blame onto Chad. But there was information relevant to your forecast, known by The Cheese, that didn't get to you, and now you just got burned by it. You don't need to take all the blame. And, of course, you need to tighten that process, which is part of your discussion with Chad.

After discussing the assumptions, you can discuss the numbers if necessary. Either the person will agree with the forecast after hearing the assumptions, or you will disagree with them, which is fine. But you will have competently defended your forecast without engaging in an argument. Again, you may have to concede that the forecast could be off due to new information.

The New Information Factor

You must first verify the information if you are presented with new information that could affect your forecast's accuracy. In the example above, you call Chad and get the details. If it is customer-oriented, contact the salesperson to determine if it is customer-specific or an industry trend.

Often, when a customer challenged my forecast, they would claim that "this" was happening everywhere, which could lead to *this*, then *this* could happen, which would totally crater demand. On these occasions, I would check with my sources and often determine that the "trigger" for the challenge was an isolated incident, highly unlikely to impact demand in the future.

If the new information is valid and changes the forecast, you should reissue the forecast with a note explaining why. And, of course, amend the forecast assumptions.

When Things Go Wrong

Even with all the transparency, communication, and effective processes, things can still slide off the rails because this is, of course, forecasting. If it weren't a difficult, roller-coaster job, you wouldn't need this book!

Sometimes this happens:

"I think the forecast is wrong!"

After you ask about the assumptions and ask their reasoning, you get this:

"There's no reason; I just think it's way too high, and there is no way we can get there!"

Well, they really haven't given much to discuss, have they? You can restate the assumptions, but they are likely trying to bait you into an argument for whatever reason. Do not get into a shouting match about the numbers.

"Well, are you going to change it?"

No, because the forecast is based on a process and assumptions, and you don't forecast on opinions, guesses, or gut-feel. You will tell the person the forecast stands based on the current assumptions. Unless, of course, it is The Cheese. This is the classic case where you must stand your ground.

The Cheese Factor

On rare occasions, the demand forecast becomes a political battle between departments. (The financial forecast is inherently political and will be discussed later.) Sales, Manufacturing, Marketing, and Finance may all enter the fray. Of course, this is all overblown, but company political infighting is expected at times — maybe continually — in toxic corporate environments.

Political fights inevitably attract The Cheese's attention, which is never good for you. Yes, now there is this perceived "big problem" with the forecast, and you do the forecast, so you must have messed up.

If this results in discussions directly with The Cheese, or if The Cheese questions your forecast independently, still attempt to discuss the assumptions they disagree with. These discussions can be tricky, especially if The Cheese is not the sharpest cheddar on the charcuterie board. Be respectful and polite as you attempt to stand your ground.

This is where the *Control* aspect of being a Master Forecaster comes into play. Everything built into your forecasting process is designed to help you maintain control of the forecast. Nothing but bad outcomes result once things spin out of control, especially in a hyper-political organization. Suddenly, there is a corporate hurricane, and you are in the eye of the storm. Everyone gets out their pistols, and looks for who is already wearing a target. Try to calm the storm by steering the ship back to a discussion of the assumptions.

If You Must Change the Forecast

If The Cheese, or your boss, after Cheese discussions, tells you to change the forecast, of course, you must comply. Standing your ground doesn't mean you make stupid, career-threatening decisions. Even the best armies sometimes retreat to fight another day. Don't resist changing the forecast if there is new information or if there is a different consensus on the assumptions.

If it is just made on a Cheesy whim, then you must still change the forecast. I didn't like it when this happened because it is no longer your forecast. Regardless, you will still measure against this number with whatever asterisks and notations you wish to document.

Hopefully, the changes are only for one or two product families and are simple to make. If the change is to the total, cumulative number, making the changes is much more complicated, and the forecast may not appear logical. You may have to discuss the changes with your primary contact in manufacturing so you can agree on the best way to adjust the production plan.

Also, the forecast assumptions will have to be changed. Make the change on your report. Leave the original assumptions in red and list the new or changed ones. Later, you will want to look at these as part of your review process to see which version turned out to be correct. There is no need to reissue the Forecast Assumptions Report; as mentioned previously, include a note on the revised assumptions with the revised forecast.

Defending The Forecast – After the Fact

Let's say The Cheese has visited your large plant in Topeka. The factory has had some production issues, and the March metrics were particularly poor. In the litany of reasons causing this, the beleaguered plant manager says, "And the March forecast was wrong!" Of course, meaning: *The forecast was not perfect.* Plant managers have a difficult job but will enthusiastically throw your geeky, corporate-cushy job keister under the proverbial bus at every opportunity. As previously noted, anyone in the company will take a shot at you if it benefits them. You have a target on you.

On Monday morning, without warning, you have The Cheese scowling at you, asking why the March forecast was so screwed up. In a good case scenario, the forecast in question was within the accuracy standard. You quickly grab the Forecast Accuracy Report, (which you will have printed out for just this occasion) then explain your process for producing solid forecasts and measuring accuracy afterward.

You can then continue discussing the forecast, pointing out assumptions that were missed that could have pushed the accuracy percentage close to the boundary. Once The Cheese has asked all the questions and has accepted all your answers, SHUT UP. Use one of the "golden rules" of salesmanship here. Once you get to "Yes", shut up — nothing you can say after this point benefits you. Once The Cheese "buys" your explanation, shut up, thank him for his interest, and offer him the copy of The Forecast accuracy report you just discussed.

If The Cheese asks why the plant had problems with this forecast, kindly direct them to Production Planning or whoever gets your forecast first. However, be professional and don't throw them under the bus.

"I don't know why that happened. I submit my forecast to Production Planning; maybe you should check with them." Keep it impersonal; you are deflecting, not accusing. Do not say, "Check with Greg Myers; he gets my forecast first". You hope that your explanation has been sufficient and that The Cheese loses interest at this point.

Thank The Cheese for their interest and invite them back anytime they have a forecasting question. You will have displayed that you are a competent, credible, confident Master Forecaster who is in control. You have stood your ground. You have turned a losing situation into a win.

Consequences of Having No Process

If you have doubts about whether you need everything recommended by my process, consider how you would have now handled the abovementioned situation. When The Cheese or your boss blindsides you, you stammer to explain that your forecast "wasn't really all that bad" and then try to throw somebody under the bus instead. You are on the total defensive, and you cannot win the argument. The arrow from the Plant Manager sticks firmly in the target. Take enough arrows, and you will be escorted to the door, even if your forecasts overall were good.

The Bad Case Scenario

In the scenario above, let's say the forecast was out of standard, and the Plant Manager's complaint is valid. (Remember, your job is to provide the most accurate forecast possible. Production's job is to take that forecast and make it work as best as possible). You still start with the Forecast Accuracy Report and show The Cheese that you know your forecast was off and by exactly how much. You then go into the faulty assumptions and a review of current market conditions. If necessary, describe what you are doing to improve the forecast process.

Point out that either you have been forecasting that product family well, up to now, or discuss why that particular product family has been creating some forecasting challenges. Answer all the questions, but now don't offer a copy of the report in this instance since the forecast accuracy will be marked red instead of green, and you don't want to emphasize that point. You will have at least presented yourself as a credible, confident forecaster, much in control of the process, who seeks to be more competent. You will have stood your ground as well as you can under difficult circumstances.

When Your Boss Is Not Happy

There will be those times when someone will complain to The Cheese about the forecast, who, in turn, will tell your boss. They may even scorch your boss because that's what The

Cheese like to do. The boss may scorch you, because the corporate crap runs downhill.

If it is a legitimate inquiry, explain the gap in the process and how you intend to plug it, if necessary. Your boss should have received your Forecast Accuracy Report before this, but he probably didn't read it, even after The Cheese discussion. Try to get your boss reoriented with the process and explain how you are working to improve it. Remind him that you measure the forecast accuracy monthly and distribute a report on it.

Nothing you can say will make things better if this is a "torch" session. If you argue, it can extend the beating and make matters worse. Simply explain what happened and shut up. The boss will then scold you some more and dismiss you back to your desk with the edict to "do better forecasts!" This is why, depending on your situation, it is sometimes better to have the opportunity to defend the forecast directly to The Cheese.

Set To Play Defense

The forecasting process in this book moves the target from your back to the front. It provides the tools to effectively blunt the attacks that will primarily now come from the front, but occasionally from the back – by the real corporate jerks. Unfortunately, even though the process will reduce the number of attacks, it cannot eliminate them. Such is the nature of the forecasting job within a politically drenched corporation. Build a strong fort around the forecasting function, gird your forecasting loins, keep your stick on the ice, and stand your ground.

For the Business Predictors

By the nature of your position, occasionally, you will swing and miss. If that happens with the bases loaded and two outs, you must explain what happened. Keep the discussions fact-based. If you documented assumptions, then discuss what assumptions were incorrect and how they impacted the outcome. You must turn the discussion from how bad your recommendation/prediction was to why things didn't turn out as planned.

Relevant Story #1

Ask questions before launching into a defense ...

Remember that story from Chapter 3 when I called the industry forecasting firm to point out his math error? That is a prime example of how not to defend your forecast. His response was much too defensive. Not only did I find it humorous that he misunderstood the reason for the call, but I also lost respect for his company because his defense of their forecast was so weak.

Bottom Line: Ask questions at the beginning of the discussion to get to the root cause of the issue. This is an excellent rule to follow in life when someone challenges your work or is upset with you.

Relevant Article #1

Here is a Whitepaper I wrote at my final job on defending your forecast to the C-suite (or the Cheese suite, if you will). Much of this article has been covered so far. Therefore, consider it a review of the material stated slightly differently. It is also specific to the trucking industry but naturally applies to all manufacturing environments.

Defending to the C-Suite: Communicating and Presenting Your Forecast to The Executive Staff

You've just finished your monthly forecast and sent that spreadsheet off to all your internal customers throughout the corporation. While that is always a welcome moment of relief, your forecast is incomplete. You need to communicate the factors on which the forecast is based to the executive staff and other interested company stakeholders.

Forecasts are never "wrong," but they always vary in their degree of accuracy. Forecasts should be based on reasonable expectations of future events. If these events happen as expected, the forecast will be highly accurate. If the circumstances play out differently, the forecast will be inaccurate. We call these expected events "assumptions." Therefore, while the forecast is never "wrong," the assumptions the forecast is based on can be wrong. So, it is essential to document and communicate those assumptions monthly in a Forecast Assumptions Report.

Reasons To Develop a Forecast Assumptions Report -

Support/Defend/Improve/Communicate

To Support Your Forecast

The forecast is based on assumptions that need to be explicitly stated. This is the "documentary" that accompanies the numbers. It is the story of how you arrived at this destination. It explains how you came to your conclusions. After reading the report, management and stakeholders should understand the logic behind the forecast and be able to explain it to others.

To Defend Your Forecast

Just because people understand the logic behind the forecast, it doesn't mean everyone will agree with the forecast. The forecast can be challenged for reasons that are legitimate, political, or just because someone is in a bad mood. Remember: the forecast is not "wrong," but your assumptions could be.

The first question to ask someone (very calmly) is: "Have you read the assumptions report?" If not, ask them to read the report and then return to discuss it in more detail. If they have read the report, ask them what assumptions they disagree with. If these are assumptions based on calculations, check and re-run the numbers. If they are based on gathered information, re-check the sources.

For example: A large order from Rogers Inc. was planned to go into effect next month. It is now expected to be two months later than planned. If your assumptions change, update the forecast as needed and reissue.

Whenever possible, you should always discuss the assumptions before you discuss the numbers. If a disagreement arises between individuals or groups, it is often because the assumptions of the most likely outcome are very different. If you address these issues early, you will find that getting buy-in from critical stakeholders is much easier.

There will also be cases where the number will be commanded down to you. In this case, it is advisable to revise your assumptions to fit with the revised forecast.

To Improve Your Forecasting Process

When a forecast is outside the acceptable accuracy standards (i.e. a "bad" forecast) it is essential you know why that occurred. Remember, the forecast is not "wrong," the assumptions are. You should review and research the results of these assumptions before anyone asks you about the forecast results. This way, you will have an assumption-based answer ready. If you are asked before your analysis is complete, you state that you are reviewing the forecast assumptions to determine the cause of the variance.

More importantly, you review the assumptions to determine the gaps in your forecasting process. This goes beyond identifying "what" assumptions were wrong and involves determining "why" they were wrong. By doing this exercise every month, you are engaging in a method of continuous improvement that will reduce the amount of "forecasting errors" and thus improve your forecasting accuracy over time.

For example: There was a big order pending in the southeast sales territory that you did not know about. Your assumption, based on the information and data that you had, was that sales in the southeast territory would be fairly stable during the forecast period. (Note: This is an example of an assumption that is not explicitly stated in the report but is presumed). You determine that the salesperson for the territory neglected to turn in a sales report last month, and the sales manager, who is expected to alert you of pending large orders, did not. You will want to have discussions with these individuals to remind them why it is vital that you have this information before you develop the forecast.

To Communicate Current Business Conditions Throughout the Organization

To develop the Forecast Assumptions Report, you must gather and report on many phases of your organization's business activity. In essence, you are providing an update on current business conditions and communicating that throughout the organization. This information will interest people who have not reviewed or have limited interest in the actual forecast.

Distributing this information to important officers and stakeholders also helps eliminate "surprises" or people being blindsided by unexpected information. You still may be the "messenger" of bad news, but you don't want people to be upset at your forecast and then be surprised by the reasons. You still may be questioned about factors in the report, but you want to discuss the assumptions and can refer the inquisitor back to the source of the assumption.

Forecast Assumptions Report Content - Economy/Industry/Competition/Internal

1. Macro-Economic Environment This includes all data about the macroeconomic environment in the markets you compete in. Determining which macroeconomic indicators and statistics are most relevant to your business is beneficial. In addition, note what indicators are mentioned most by the executives in your organization. You can always ask people which data is most relevant. Macroeconomic analysis and data are in every report put out by the company.

2. Industry Analysis

What factors in your industry (or industries) will impact your forecast? The company gives you a comprehensive, expert analysis of the commercial transportation and freight markets. For other industries, the process is essentially the same. Break the section down into three parts:

a. (Trucking) Environment

What are the significant issues impacting trucking in general? This information is provided in each Truck & Trailer Outlook report and the database and graphs package.

b. (Truck) Equipment Analysis

What are the factors affecting equipment orders and shipments? Once again, the company provides this information directly within the Truck & Trailer Outlook service.

c. (Truck) Equipment Forecast

Present the company forecast here in whatever detail management wants to see it. If you are modifying the forecast based on your analysis, state why. The equipment forecasts are presented in tables within the Truck & Trailer Outlook report, in the accompanying charts package, and within a database with a full dataset of history and forecasts.

3. Competitive Environment

Here, you list the assumptions regarding competitive actions. Include information on special sales and marketing campaigns, pricing strategies, product introductions or quality issues, and promotional activities. You get this information from sales reports, marketing updates, and sales and marketing personnel conversations.

4. Company Environment

These are the assumptions that are unique to your company. You review sales and order trends. You discuss potential orders that impact the forecast. You list price, product, and promotional issues that will impact sales in the forecast period.

Note: You should also list any production or supplier issues that may impact the forecast. In addition, any issues that have an impact on forecasted market share should be stated.

5. Conclusions and Summary

This summarizes what changes were made to the forecast based on all the assumptions. The first paragraph should summarize the key factors influencing the forecast this month. The second paragraph starts with something similar: "Based on these factors, the forecast was increased 7% total over the next 6 months…"

Forecast Assumptions Report Format & Timing

It is best to list 4-6 bullet points per section. Management does not have time to read long blocks of copy. Also, this is not an analysis of all the issues but a list of assumptions that support your forecast numbers.

Include graphs to enhance the appearance and provide context. You can rotate the charts monthly to make the report more dynamic and fresher for its audience. Unfortunately, you will have to distribute your forecast before your assumptions report.

Your internal customers will need the numbers to complete their assignments. Because the assumptions support the forecast, you will want to complete and distribute the report as soon as possible (this is the next step in your forecasting process). If you are asked to defend the forecast before the assumptions report is released, ask for more time to complete the report.

Relevant Article #2

This essay originally appeared in my fourth book, Deep, Heavy Stuff, which is about the problematic issues in life. However, since we as forecasters and business analysts, are continually having our work questioned, I have included it here. Those challenging us in a work environment are often arrogant, rude, and obnoxious. If we react poorly, it can damage our careers or end our current employment. This may be one of the most important sections of the book. I have modified the essay for business situations but the concept obviously pertains to all personal relationships.

React or Respond? – There is a vast difference.

Your boss, coworker, customer, or vendor can unleash a verbal tirade upon you unexpectedly. Or, it's just a casual statement that highly annoys you.

You feel (pick one or many) angered, threatened, demeaned, disrespected, marginalized, insulted, offended, disgusted, fearful, irritated, or repulsed.

So, how do you react? Or -- how do you respond?

Don, that question is redundant, isn't it? React and respond are the same thing, right?

No, they are not. And understanding the difference can change your life.

Are you telling me that knowing the distinction between these two similar words is that important? How?

When you react to people, it is an emotional outburst, most likely delivered with the same antagonism as the offending statement. You speak without thinking; the words fired out of your mouth like a submachine gun. They are angry, sarcastic, venomous, biting, cutting, and hurtful words.

Reactions can destroy relationships and reputations. Reactions can produce hurt feelings and pain that sometimes take years to heal. Reactions cost employees their jobs, promotions, or company reputations.

Reactions almost always lead to arguments rather than discussions. Unfortunately, the arguments lead to even more reactions, with both parties' sub-machine guns blasting away. If you need examples, look at the politicians, twitter wars, or the flaming fights on Facebook.

Well, yes, reactions are bad, but why are responses any different?

A response is a rational, calmer action designed to diffuse the situation and start a discussion instead of a heated argument. An essential aspect of a response is understanding why the person has just unloaded upon you. Therefore, one of the best responses starts with you asking a question. If you can't think of a question related to the subject, a good option is: Why are you so upset?

This allows you to understand the situation better, enables the other person to continue to vent, and gives you time to formulate a second, more relevant question. Sometimes, the "Why are you angry?" question will change the discussion's tenor as the person realizes they have spoken too forcibly. Often, they will apologize, and then a purposeful conversation can begin.

But how does letting the other person keep ranting and raving help things?

Because it puts you in control of the situation. While the other person is out of control, you can formulate your response. You may choose to disagree, but it is possible to do so in a calm, mature manner. You may even decide to be charming in your reply. But the goal is to defuse the situation, not escalate it. Unfortunately, your calm response sometimes enrages the person even more because they are looking for a fight and will start spouting off again. But then, still, you can remain in control of the situation and respond accordingly. Just because they want a fight doesn't mean you must participate.

There are situations in which you will have time before you choose to respond or react. For example, when you get upset by an email, post, or tweet. How often have we reacted to that situation with a nasty missive back, fueled with emotion? How did that work out for you? It made you feel good in the moment – caused you regret for a much longer time. When you have time to formulate a response, the critical question is: What do I want to happen next?

Once you know what a desirable outcome of the situation is, craft a careful, strategic response that supports your intent. Resist the temptation to say to yourself: "I'm going to give him a piece of my mind!" "I'm going to show her just how stupid she is." "I'm going to show them who's boss," etc. etc. etc.

These reactions make us feel good in the moment, but if after the smoke clears, the situation hasn't changed, or maybe now it is worse, what have you gained? You still wake up tomorrow with an ongoing conflict or problem.

Look! There's a raging conflict burning out of control. Is it better to pour gasoline on it – a reaction? Or is it better to pour water on it – a response? You very rarely have to apologize for a response, but you often have to apologize for a reaction.

Okay, reactions are harmful. But they are just natural tendencies. It's what I do in that situation. It's what I've always done. How do I change that?

And now, we come to the difficult part of this post. Easy to say, hard to do. This is a learned behavior, but the sooner you understand and can implement this change, the better your life will be. Again, it can improve your career and business relationships. It is one of the most valuable life skills to acquire and practice.

So, the next time you are faced with a conflict situation: Will you react, or will you respond?

CHAPTER 16

"Meeting" Expectations – The Forecast Meeting

An essential part of forecasting is formally meeting with your internal customers to discuss the forecast and Manufacturing's response. The ideal time for the meeting is after all the information has been collected and the data analyzed and right before the forecast is finalized.

Sales & Operations Planning – S&OP Meeting

The Sales and Operations Planning (S&OP) meeting is the gold standard for coordinating demand and supply plans in an organization.

This monthly meeting is led by senior management to coordinate the demand and supply plans with the business plan. The goal is to gain consensus on an operating plan that optimizes company profits. The planning period is usually 12-18 months.

The standard meeting agenda includes Product Review, Demand Review, Supply Review, Finance Review, and S&OP discussion, with the final S&OP plan developed after the meeting.

I never participated in a standard S&OP meeting, so I can't provide any insight here. At a previous employer, The Big Cheese had a brilliant business mind, and most long-term planning was not done in a meeting but between his ears. However, most companies don't have a Big Cheese that astute, therefore long-term planning by having S&OP meetings is recommended. If you want information on official S&OP stuff, there are plenty of books, seminars, consultants, and software available.

Forecast Meeting

An alternative to a full-blown S&OP meeting is a monthly forecast meeting. This meeting is more short-term in nature. The discussion is focused on demand 3-6 months out, with the emphasis, of course, on the key forecast month.

Relevant Story #1

If you have a forecasting team, this is your huddle ...

Forecasting should not be conducted in a vacuum. There must be coordination, communication, consensus, alignment, cooperation, and teamwork. All relevant parties need to meet monthly to discuss the forecast.

After all the other pieces of the forecasting process were in place, I held my first Forecast Meeting. Because of the prior disastrous forecast meetings at a previous company, I was extremely apprehensive. I invited the Materials Manager and his two key planners (per his request), the VP of Operations (his boss), the VP of Marketing (my boss), and the Customer Service Manager. The Sales Manager, who worked remotely, was patched in by phone.

I made the risky decision not to invite the Big Cheese to the meeting. As described earlier, the previous forecasting meetings were the worst business meetings of my career. In the prior meetings, this Cheese would start off by berating everyone and not stop until he ran out of breath – "the beatings will continue until morale improves" approach. Did I really want that guy in a forecast meeting I was leading? That would be like putting a huge flashing target on my face and wearing a dunce hat. Yes, I took a big risk in excluding him, but it was the correct choice.

Of course, he was always welcome to attend if he found out about it. If asked why we were meeting, I would have used my standard line when asked about any of my forecasting process changes: "This is how the experts say to do it." I used this statement to thwart any resistance to the forecasting changes I was making at the company.

In the new forecasting process, I implemented the Forecasting Meeting last since, essentially, it meant I was going "public". Up to this point, I had quietly built the process. Still, in holding this meeting, I was proclaiming: "I'm the person in charge of forecasting. This is my forecasting process. This is my forecast. And if you have any questions, please ask." I was now taking full responsibility for the forecast.

The Forecast Meetings were a huge success. The VPs only attended occasionally when they wanted an update on the market or if issues arose. The Sales Manager called in if we had specific customer questions. I don't know if the Big Cheese ever knew of the meeting, but he never attended – which benefitted everyone.

I knew the meetings were successful when the Purchasing Manager asked if he could attend, which he could, of course. No one wanted to go to the old forecast meetings; now, people wanted to come to mine. The old forecast meetings were the worst of my career, but the new ones were some of the best.

Forecasting Meeting Format

As previously mentioned, the format of these meetings is highly flexible depending on your company and your industry. The following is how I conducted my meetings. Feel free to modify and improve upon my format. (Continuous improvement rules!)

1. **A discussion of the macro-economy.**

   Present the charts and graphs of the key economic indicators you are tracking which are currently relevant. The primary indicators will be displayed monthly. Various other indicators can be rotated in so the presentation has some variety.

   This gives the group a shared view of the economic environment. You will expand your perspective as your team comments on the indicators. This prevents you from operating in a vacuum, missing key details, and making wrong assumptions. People will raise new issues and add relevant viewpoints. There will be suggestions about additional indicators to discuss at future meetings.

2. **A discussion of the industry/market.**

   Next, discuss the industry/market conditions and all trends that may impact the forecast. Show the relevant charts and graphs for the month.

   Again, the discussions can reveal new perspectives or theories for future examination. Someone in the meeting may have heard something relevant to the forecast.

3. **A discussion of the customer environment.**

   Report on what is happening with key customers that will impact the forecast. Sales and Customer Service can report what they know.

   This provides a forum for Manufacturing and Sales to interact. The information exchanged here can even be longer-term in nature. For example, Manufacturing will appreciate the heads-up if Customer B plans to open a new plant in Texas eight months from now.

4. **A discussion of the manufacturing environment.**

   Manufacturing reports on its efforts to meet customer demand. Issues created by the forecast are examined, and supply chain issues are highlighted. This is an opportunity for Manufacturing to express its needs and issues related to the forecast for the next few months.

   Since Customer Service and Sales are present, be aware that the discussion may veer off course and shift to why specific delivery dates were missed. While these discussions are relevant, they are beyond the scope of this meeting. If an argument arises, suggest that the participants meet later to continue the conversation.

5. **A review of the key product families.**

   Discuss what is happening with the important products regarding market share, trends, and competitive threats. If Marketing is present, they can give a brief overview.

6. **Current forecast review by product family.**

   Show last month's forecast for the key forecast month. If May is your key forecast month for the meeting, then show the forecast for May from last month to establish a baseline.

   The goal of the discussion is not to establish a forecast number but to attempt to gain some consensus on DIRECTION. Is demand increasing, decreasing, or staying the same? Does last month's forecast look reasonable, or does it need an adjustment?

   In many cases, reviewing all the factors will generate some consensus of opinion. When the forecasting picture is cloudy, divergent opinions will exist, but healthy debate is beneficial. You don't need complete agreement. You need the interested parties to express their views so that you can consider the various viewpoints when constructing the final forecast.

7. **Summary and general statement on probable forecast adjustments.**

   Summarize the discussion and give the group an indication of the direction of the final forecast. Answer any remaining questions.

Benefits of a Monthly Forecast Meeting

1. **It fosters teamwork and comradery.**

   Bringing the "forecasting team" together enhances the buy-in to the process and the final numbers. When you win, you win as a team. Your company benefits, and there is nothing The Cheese likes better than interdepartmental teamwork. We have already seen what happens when everyone is upset and blaming the forecast.

It is dangerous when you are isolated in a "forecasting silo", emitting the numbers like a puff of smoke indicating the selection of a new Pope. If you go it alone, you are wide open for attacks coming from all directions. If you can establish a "team forecasting" environment, your cohorts will defend you from criticism, even when you are not present. This is invaluable to you in flourishing in your job and remaining employed. It is much easier to stand your ground with a team providing support behind you.

2. **It provides transparency and "no surprises".**

Based on the discussions and after fully disclosing the data and information, the final forecast numbers should meet expectations and provide no surprises. There should not be any potshots or drama when the numbers are distributed, by anyone in the meeting. (Of course, we have special names for people who don't speak up in the meetings but howl like wolves afterward.)

This doesn't mean everyone will agree with the forecast. It isn't ever easy to get a complete consensus. However, you can usually generate a "general agreement", and often, questions later raised by people who were not in the meeting can be answered by those who were in the meeting.

3. **You benefit from the "wisdom of the crowd."**

   Now, given that five or six people in the forecast meeting is not a huge "crowd," these people are hopefully intelligent and motivated to produce an accurate forecast. Sometimes, you get so immersed in the pile of data and information you have analyzed that you miss the obvious. Other times, someone will see something from their unique perspective, leading to a productive discussion. Sometimes, someone will ask a seemingly stupid question that triggers your brain into a eureka moment.

   There are also those cases, hopefully infrequently, where you are wrong, and the team must steer you back on course. However, sometimes you must "stand alone," as we will cover in a later chapter.

4. **Everyone has an opportunity to contribute and learn.**

   During the meeting, everyone can express their opinion and contribute to the process, giving people a sense of ownership and value. In addition, people in the meeting gain knowledge about the company and industry, which will benefit them in their current and future positions.

Other Details

When demand fluctuates, and there is market variability, the meeting can take two hours. When there is stability, the meeting could end shortly after an hour. You must focus on the forecast and the current outlook, with little discussion of the past.

On that note, I don't recommend reviewing the forecast accuracy during this meeting. There is just something about putting those charts on the screen that can cause the discussion to careen into places you don't want to go, especially if your boss is in the room. The production planners, who care most about your performance, have already analyzed the charts because they need to consider it when developing their next plans.

The production planners were keenly aware of my forecasting biases. Every month in the meeting, they heard my logic and discussed the upcoming forecast. They got to pick my brain for more details if they needed additional information.

After the meeting, they probably discussed my biases, something like this: "I think Don is overestimating how fast demand is going to recover." Or, "Don's forecast is just not catching up with the increased sales of Product C."

These observations are expected because your forecast is not perfect. Your job is to provide the most accurate forecast possible. Their job is to develop the best production plan possible based on your forecast. If the production planners do their job well, other people will seldom criticize your forecasts. So, do not take skilled production planners for granted—they can keep the arrows from flying your way and help you stand your ground.

Maintaining Control

You are the chief forecaster, and you own the forecast. Therefore, you should run the forecast meeting and publish an agenda based on the abovementioned steps.

Because the forecast can be such a political/volatile topic, you must maintain control of the meeting. The discussion can veer off course at any moment. Sometimes, those "side" discussions are unrelated to the forecast. Some participants may not have discussed essential issues or conflicts that have recently arisen and now want to hijack your meeting. Suggest they discuss the issue outside of the meeting. Of course, if some Cheese is in the side discussion argument, consider your meeting hijacked and wait for an opportunity to get things back on track.

An essential factor in maintaining control is knowing what you think the general forecast should be before starting the meeting. By that time, you will have analyzed all the data and information. You could logically already complete a final forecast, but it is vital to have this meeting for the reasons previously stated.

Use your anticipated forecast number as the beginning "marker" and let the discussion proceed, up or down from there. For example, when reviewing the previous 4-month forecast for Product Family D, you might say, "Based on the information we've discussed so far, it would appear that last month's forecast should be raised." Versus, "Well, what do you think the forecast for Product D should be?"

If you do not do this, the discussion can jump the tracks, and you can end up talking about wild numbers instead of forecast assumptions. This can lead to long, meaningless debates, waste valuable meeting time, and derail reaching a general consensus.

Guide the discussion where you believe it should go, but be flexible enough to change course as needed. Bottom line: Be in control of the meeting and the conversation.

Relevant Story #2

Always throw out an "opening bid" ...

At my final employer, I was in charge of the forecasting process for commercial vehicles (three distinct sectors). The process here was different because we forecasted industry demand/production rather than individual company demand.

The process was meticulous, usually taking a team of six people working over 40 man-hours. Our economist led the Economic Meeting, during which the economic forecasts were plugged into our model to provide freight data. Then, our freight expert led the Freight Meeting, during which the freight forecasts were plugged into our demand model to provide the baseline for our equipment forecasts.

In the Equipment Forecast Meeting, I reviewed all the market metrics, such as order rates, backlogs, sales, cancellations, etc. But when it came time to discuss the forecast, I learned to put out an initial number to start the discussion. If I didn't and let this band of six highly intelligent, opinioned experts go at it, we could debate the numbers for hours. So, I put out the opening bid without commentary, which everyone understood was my recommendation, and then the debate began.

This method saved us a lot of time. Sometimes, we would spend the last five minutes of the meeting arguing over relatively small changes in the final number—a difference irrelevant to our customers. But that spirited debate and attention to detail made us Master Forecasters.

CHAPTER 17

Putting It All Together – The Monthly Forecasting Process

Like many other concepts we have covered, your forecasting schedule depends on your company's schedule. The primary deadline is usually when Production needs the demand forecast to begin their planning cycle. You can negotiate this, but it will come down to whether you adapt to their cycle or they adapt to yours. And when push comes to shove, they are your internal customers, so follow their schedule as needed.

The steps are listed chronologically, starting with the first of the month. This is a continuous cycle, so it really doesn't have a starting or ending point.

Step 1 (Beginning of the month)

Accuracy Measurement

Let's say it's early May. You can download April's production data as soon as it's available. Then, plug the product family numbers into your accuracy measurement spreadsheet. The spreadsheet is set up to measure against your key forecast month. If you are measuring against a 3-month forecast, you would be comparing April's actual result versus the April forecast you published in January.

Each product family's results are shown in a bar chart. Green bars are within the accuracy standard (that you established), and red bars are your misses. I would hit 9 out of 12 product families in a good month. If 8 out of 12, it was mediocre. Anything under eight was a weak forecasting month, but when I hit 100%, I strutted around the office that day.

The results are the results, and the standards are the standards. If you crawl under the mark by .01%, it's a win, but exceed it by the same amount, and it's a loss. And a tie goes to the runner because you set up the system, so you get to determine the rules!

You can make exceptions for catastrophic events. For the COVID-19 months, I recorded the percentage miss but colored the bars blue on my internal charts to designate a jumbled environment. However, you should measure the results regardless of the circumstances. At my final employer, we compiled a forecast accuracy report for 2020, available to any customer or potential customer who asked.

Forecast Accuracy Analysis

Now, you analyze the product families that were out of standard to determine the reasons. You dive deeper into the numbers to see if there were changes in customers or specific product sales. Misses in the smaller product families, especially those with fluctuating demand, are often easily explained.

Missing the standard on the more important product families can be more challenging to explain. If you were tracking April sales during the month, you may have already known that your forecast was inaccurate and started looking for clues.

You will often be alerted as soon as someone deems the forecast to be *wrong*.

Best Case Scenario: The production planner calls and says, "Hey, just a heads up. The April forecast is off. We are running way behind on orders."

Bad Case Scenario: Your boss summons you and says, "They said in the staff meeting there was a problem with your forecast."

Worst Case Scenario: The Cheese stands over your desk and barks, "You mucked up the forecast!"

The customer analysis can reveal specific shortfalls or overages to the forecast, possibly necessitating a follow-up call to the salesperson. You should keep these conversations positive since you need the salespeople to provide information to you regularly. Often, the salesperson will appreciate the heads-up since they may not yet have had a chance to review last month's sales, which allows them to research this before their boss calls about it.

Product sales fluctuations may lead to discussions with product managers and salespeople if the differences are isolated to a sales region. The January forecast assumptions for the April forecast need to be reviewed for validity.

Issue the Forecast Accuracy Report

You now have the information needed to issue the forecast accuracy report. Include all the bar charts tracking the forecast accuracy for the key month against the accuracy standard.

Only a few comments are needed for those product families that are within the standard. Point out when the forecast was remarkably accurate since you need to broadcast your successes to build political capital and offset your misses.

For the product families outside the standard, list the reasons for this. A key element here is that you are documenting **reasons** the forecast was outside the accuracy standards, not **excuses**. Do not say, "The forecast was too high because sales failed to tell me about the competitor's promotion."

You are providing logical reasons for an inaccurate forecast, but this does not absolve you from the results. If there is information you didn't know, why didn't you know it? Faulty assumptions were bad assumptions, although this goes back to your ability to see and predict the future.

Of course, there are those events that you can't predict that knock the forecast out of standard. The most frustrating case is when your forecast would have been 14% off versus a 15% standard, but a small circumstance, insignificant to anyone else, bumps your forecast accuracy out to 16%. It's labeled a miss, but explaining how you should have "won" but lost can be difficult.

You send out the report. The first time it goes out or the first time it is read, you will receive questions about the process, standards, etc. After that, it will usually be skimmed or ignored, except by the production planners. You may receive questions when you have a weak forecasting month. However, this should be a discussion on why the forecast was off and not a torch session on your "poor" performance.

Distributing the report is a defensive move. If the bad news is already public, people are less likely to try to blindside you for an inaccurate forecast. It feeds into the "no surprises" strategy. If a "fire" on the forecast does break out, referring back to the accuracy report is your first line of defense. Issuing the report monthly displays that you are a competent forecaster with control of the process and helps you to stand your ground.

Relevant Story

You are the forecaster – you are responsible for the forecast, not the actual results.

One time, shortly after releasing the Forecast Accuracy Report, I found The Cheese in my office with a scowl on his face. He wasn't on the distribution list for the report, but someone had referenced the report to inform him about lower sales last month. Something like "Ake's report said Southland's sales were really low last month."

Ironically, my forecast was accurate because I had already factored in a sales dip that month. But my Forecast Accuracy Report was the first to report the bad news.

Cheese: "Southland's sales were bad last month?"

Me: "Yes, sales were lower than usual."

Cheese: (scowling) "Oh! Why was that?"

The Cheese thought that, somehow, I was responsible for Southland's poor sales that month. Well, of course, I only forecast demand; I don't impact it. I took a deep breath and carefully replied:

"I don't know. That's probably a question for Jensen (the Sales Manager)."

Step 2

Conduct Statistical Analyses

Now that you have the April data, plug/link the data into the tracking/analysis spreadsheets and conduct your trend/statistical analysis. You can also calculate the more complicated forecast accuracy statistics. Review the customer sales for April for anomalies and trends. Make notes that you can include in the Forecast Assumptions report.

Step 3

Review All the Market Intelligence and Develop Assumptions

Now, review all the market information/intelligence you have gathered. Collecting the market intelligence is actually Step 10, which happens as soon as the monthly forecast is completed (remember, this is a continuous cycle). You start collecting the data in Step 10 as soon as the final forecast is released. Here, imagine opening the jigsaw puzzle box; you have all the pieces needed to create a picture; the challenge is putting them together.

By combining this information with your statistical analysis, you can clearly see the picture of future market demand. Of course, this is more difficult when conflicting information exists (the pieces don't fit together well), which often happens during a market transition.

Document the significant factors and conclusions. Based on your analysis, develop your assumptions. These assumptions are then discussed in the Forecast Meeting and documented in the Forecast Assumptions Report.

Step 4

Produce the Preliminary Forecast Numbers

Run the updated numbers through your forecasting software for the statistical results. Prepare the bottom-up forecast by estimating demand by customer for the key forecast month. Construct the top-down forecast based on total market demand and assumptions about market share and other factors.

Compare the three forecasts, review the accuracy of the three methods versus the April results and recent history, and determine preliminary forecast numbers for the key forecast month.

Step 5

Conduct the Forecast Meeting

Assemble the presentation for the Forecast Meeting, including the charts, graphs, and all other relevant information. Conduct the meeting, taking note of additional comments and information provided by the participants. Present a proposed forecast.

Step 6

Finalize, Complete, and Release the Forecast

Finalize the forecast for all the product families for the key forecast month based on the discussions in the Forecast Meeting. Fill in the rest of the months on the forecast sheet.

Assuming a six-month forecast, the recent months (months 1 and 2) are based on backlogs and current trends. Beyond the key month, the outer months are based on trends and market information.

Send out the forecast to the distribution list. The list should include all participants of the Forecast Meeting and all interested parties, such as Purchasing. Other departments, such as HR, may want a heads-up if demand is surging or plummeting.

Note: When in the month the forecast is released will influence what Manufacturing will want as the key forecast month. If you release the forecast on the eighth business day of the month, Manufacturing will use this in their lead-time calculations to determine how long they need to respond to the forecast.

The timetable presented here is the most favorable one to the forecaster. Manufacturing may require the forecast to be released at the beginning of the month to fit in with their planning cycle. If so, you must estimate (extrapolate) shipments for the month based on current data and then update your tracking spreadsheet later with the final numbers. Just be flexible and adjust the process presented here to your situation.

Step 7

Complete and distribute the Forecast Assumptions Report

Ideally, the Forecast Assumptions Report would be sent out with the forecast. However, in most cases, Production wants the forecast number as soon as possible and will review the assumptions later. And most of the assumptions were already presented in the Forecast Meeting.

This gives you more time to compile the Forecast Assumptions report. But issue it as soon as you can after releasing the forecast. Ideally, send the forecast and assumptions report together.

Use the analysis and conclusions from Step 2 and Step 3 to develop your assumptions and include any additional information from the Forecast Meeting. To streamline this step, it is important to take clear, organized notes throughout the data/information steps and during the meeting.

Step 8

Take a Breath – You've Earned It!

Grab some coffee, catch up on the latest office gossip, and celebrate if you had an excellent forecasting month. If the results are not stellar, strategize what you will do to plug those gaps, improve your process, and tighten your forecast accuracy.

Step 9

Clean Up and Catch Up

Catch up on all the work you set aside to complete your forecast on time. Delivering the forecast is always your first priority unless overridden by your boss or The Cheese. Also, clean up and update any files used to formulate the forecast.

Step 10

Begin Collecting Information and Data for Next Month's Forecast

Now, it's time to begin next month's forecasting cycle. Decide who in your group of experts you want to talk to. Research and gather relevant news articles and chat with the Sales Manager and salespeople.

Step 10 is really Step 1 in the process. It is listed here because this happens at the end of the month, and the steps are listed chronologically from the beginning of the month. It is assumed that Production wants the forecast as early in the month as possible. Adjust your schedule as needed because your conditions, timelines, and requirements may differ from this timetable.

Consider this your 10-Step Program for better forecasting. Follow this plan, and you will attain the rank of Master Forecaster! And, of course, the process enables you to stand your ground.

For The Business Predictors

Of course, your process will look much different depending on your task. Document the steps in the process and repeat them every time you conduct an analysis. Consistency is the key to success. Develop a checklist because the most significant errors in business usually happen when vital steps are skipped.

CHAPTER 18

Miscellaneous Forecasting Issues

Long-Term Forecasting

A request will be made periodically, perhaps yearly, to produce a long-term forecast, typically five years in length. The approach here is similar to the near-term forecast but expanded somewhat since you usually have more time to complete it.

It is best to have a long-range, S&OP-type meeting to discuss demand factors for the next five years. The Manufacturing Cheese should attend to evaluate future capacity requirements.

Formulating a base 5-year forecast before holding this meeting is beneficial. This provides a starting point for the discussion rather than leaving it wide open for debate.

If you don't, the meeting can derail into a rambling discourse between sales and marketing about their dreams for the next five years. There will undoubtedly be a debate between sales and marketing over strategy and direction.

If a Big Cheese is in the meeting, there can be even more blustering. If the meeting is hijacked, you will have wasted two hours of everyone's time. The Big Cheese may determine that so little has been accomplished that still another meeting is required, which risks wasting additional time. Even worse, The Cheese may assign you forecasting tasks, essentially telling you how to do the forecast. This means you have lost control of the forecasting process, which is never good for you.

Instead, start by collecting all of the 5-year economic forecasts available. Yes, these forecasts are speculative, but so is your long-term forecast. You need base assumptions; in almost all industries, it starts with the general economy. Next, talk to the big-thinker, big-picture contacts on your expert list to get their impression of where the industry is going in the next five years.

You will want an in-depth discussion with Marketing about new product plans and forecasts and current product projections over the forecast period. Also, get their outlook on market opportunities and threats – the back half of a SWOT analysis. You also need to have a similar discussion with Sales about what trends they see and the opportunities and threats. You can ask them about the sales potential for the new products Marketing has planned. And you hope you don't hear, "Nah, that's not going to sell. What we really need is ...."

Getting To the Numbers

1. Run the historic numbers through your forecasting software to generate your 5-year baseline. This assumes current trends will continue throughout the forecast period. If ever asked to provide a 5-year forecast quickly (like being asked in the morning to have it that afternoon), run the statistical forecast and adjust it based on any additional industry data you have.

2. Create a spreadsheet to document all your forecast assumptions and their predicted impact. The assumptions run down the rows on the left, and the individual years are the columns.

    There will be assumptions about the economy and industry and their impact on the forecast. List new products separately with their forecast per year.

Assumptions about existing products and their impacts are also listed in this master spreadsheet.

3. Total the numbers in each column to determine the impact of all the assumptions for each year.

4. Add (or, egad, subtract) these to the baseline number from the statistical forecast and create the initial 5-year forecast.

5. If forecasting individual product families, apply the relevant assumptions to each family and adjust the base statistical forecasts.

6. Convert the units into dollars if a financial forecast is needed. Be sure to document the pricing assumptions. This adds more variability, since predicting price changes five years in the future is especially challenging.

Now the Meeting

You can have the meeting once you have a preliminary forecast with all the documented assumptions. There will still be the same level of discussion and debate, but now you have provided something specific to debate about. The Cheese will likely challenge and change your assumptions and come up with preferred numbers. After the meeting, modify the assumptions and impact numbers to match the agreed-upon forecast and distribute the report to the attendees and interested parties.

One Final Word on This

Most 5-year forecasts are overly optimistic because all new products are expected to be winners, and market shares are always expected to grow. The good news is that 5-year forecasts are rarely reviewed, so the accuracy here matters much less. But keep those assumptions well-documented and saved, just in case. The benefit of developing the 5-year forecast is not the numbers themselves but to get The Cheese and relevant people thinking about long-term strategy and production/resource planning.

Pure Demand Forecasting

Some forecasting experts advocate doing a "pure demand" forecast. This involves forecasting the total demand for your product, regardless of whether you can produce it.

This forecasting method is most relevant when rapid growth in demand outstrips production capacity or when disruptions to your outputs are due to factory fires, labor strikes, supply chain delays, etc.

While theoretical arguments can be made for this approach, it isn't practical in the real world. "Missed demand" can be challenging to estimate. You can't rely solely on sales reports because salespeople tend to inflate the lost sales. In addition, the forecast accuracy can't be measured when you are producing at total capacity because "what you could have sold" is an estimate, not a hard number. Forecasting total demand at 1,000 units is useless and counter-productive when the factory can only produce 800. Your accuracy will be calculated at -20% but it is a meaningless statistic.

Total demand forecasts during production restrictions do have value. If demand has outstripped capacity, the forecast was much too low, or your S&OP process failed to prepare for the upswing. An accurate pure demand forecast is essential at this point in planning increases to production capacity. In rare cases, the demand surge may be temporary, and long-term expansion is unnecessary.

When production is temporarily constrained, like when there are supply chain delays, The Cheese will need a pure demand forecast to report the financial impact to the Board of Directors, Shareholders, or owners. A pure demand forecast will also be beneficial in estimating how long it might take a company to catch up with pent-up demand.

If you must provide a pure demand forecast every month when demand exceeds production capacity, then develop two parallel forecasts. Your base forecast will now be a "supply forecast". Manufacturing tells you what they expect to build in the key forecast month, which becomes your forecast. Your forecast accuracy is excellent during these times but means little. Measure it regardless, as part of the process.

Supply forecasting during supply chain shortages, as during The Great Supply Chain Clog of 2021-2022 (into 2023 for some), can be a nightmare. Your supply forecast will be based on what materials/components Purchasing expects to receive three months from now. Let's say 1,000 units of a critical back-ordered component are expected to be delivered in April. You forecast 1,000 units for the product needing that component. But surprise! Only 500 units are actually delivered, and your forecast is off 50%. Tilt!

Continue your standard forecasting process even during unusual circumstances. Always document your assumptions, like the 1,000 components on assumed delivery, and measure against the most reasonable metric available.

Why You Can't Trust Order Volumes

Under industry supply constraints, you cannot trust order volumes as a true measure of industry demand. Typically, this happens if total market demand surges beyond total industry supply and manufacturers scramble to increase output but are limited by labor and component availability or if the supply chain for the industry is clogged.

Either way, your customers, distributors, vendors, stores, etc., are desperate for products. Brand loyalty can disappear depending on the product and the severity of the shortage. This can result in customers ordering 200 units from four different suppliers, hoping to secure the needed product from any source. Depending on the current circumstances, once the customer secures the 200 units, they may cancel the orders for the other 600 units. In this case, total demand was overstated by 300%. When all customers do this in a panic situation, your backlog swells, but it is a false flag.

During supply shortages, your customers will place orders far out in the future to reserve production slots. These "placeholder" orders do not represent actual demand. They inflate your backlog and overstate potential sales. If the supply crunch is over when the placeholder orders become *current*, the quantities will be significantly reduced or even canceled.

Order cancellations, huge ones, can wreak havoc with your forecasts. Some industries don't have cancellation penalties, making matters worse. And cancellations can come quickly and in waves. At a previous employer, in the weeks after the financial crisis hit in September 2008, every order for one product line scheduled for December was canceled. Of course, I didn't hit my accuracy goal on that one!

Conclusion

Pure demand forecasting has a purpose but should not be used as a standard forecasting procedure. Many of the forecasting rules change during supply shortages and manufacturing disruptions.

Relevant Story

As previously mentioned, the Great Supply Chain Crisis of 2021-2022 hit commercial vehicle manufacturers early, permitting me to clue in The Wall Street Journal.

However, I vastly underestimated the severity of the crisis. Based on history, I thought our industry would only be impacted for a few months. Because demand in the industry experiences wide swings, there are always supply chain issues during upswings; therefore, manufacturers are adept at managing them. Of course, this was the worst supply chain clog since WWII.

For the next year, my final employer was forced to forecast what the industry could build, not industry demand. We became supply forecasters instead of demand forecasters. This was extremely challenging because our models and analyses were designed to forecast demand rather than supply. Our forecast was off at times, and the team would become dejected. I repeatedly reminded them, "We are excellent demand forecasters, not supply forecasters."

If I had anticipated that the supply chain clog would last two years, I would have tried to devise a system to forecast supply better. However, this would have remained difficult since semiconductors are a critical component in Class 8 trucks, and some of the expert forecasts for semiconductors by big-name financial firms turned out to be ludicrous. So, we did the best we could, which overall was respectable, and our customers were grateful for our work during the recovery period, despite the bizarre market conditions it created.

Relevant Article

I wrote a blog post in July 2021 to explain the current misalignment of supply and demand.

How Demand Outpaced Supply

Mr. Demand and Mr. Supply were constantly jogging along the Economic Trail. Most of the time, they ran in tandem at about the same speed. But sometimes, Mr. Supply ran ahead of Mr. Demand, and often, the roles were reversed. But then, they would instinctively adjust their pace to get back into balance once again.

Mr. Price always accompanied Mr. Demand and Mr. Supply on the trail. Mr. Price could be unstable, so he rode a bicycle instead of jogging. When Mr. Demand got ahead of Mr. Supply, Mr. Price would pedal faster to get ahead of Mr. Demand. Then, he would force Mr. Demand to slow down so Mr. Supply could catch up to him. When Mr. Supply ran ahead of Mr. Demand, Mr. Price would drop back, causing Mr. Supply to slow down and Mr. Demand to speed up. In both circumstances, as soon as Mr. Demand and Mr. Supply were back in sync, Mr. Price could be found pedaling right beside them.

Everything was progressing fine on the Economic Trail until March 2020, when Mr. Demand and Mr. Supply were infected with COVID-19. Both guys immediately stopped and sat down on the trail to recover. Fortunately, Mr. Demand had only a mild case of the virus. Soon, he felt much better and started running at a modest pace again. Unfortunately, Mr. Supply got much sicker from the virus. He lay weak and dormant for a couple of months. He finally got up and tried to run but stumbled along slowly.

Mr. Demand continued to recover and picked up the pace. His Uncle Donald even gave him some strong coffee to help his recovery. Mr. Demand was now running much faster than still sickly Mr. Supply, and the gap was widening. Mr. Price had also been infected but was asymptomatic. He had stopped on the trail, waiting for Mr. Demand and Mr. Supply to recover. He had followed Mr. Demand back on the trail and expected Mr. Supply to follow them, but he had not.

Now, Mr. Price was pedaling faster but didn't know what to do since Mr. Demand was not slowing down and wasn't concerned that Mr. Supply was falling so far behind. The gap between Mr. Demand and Mr. Supply had not been this wide for many years.

Even though Mr. Demand appeared healthy, his Uncle Joe was concerned he was not running fast enough. So, Joe gave him an energy drink, a stimulus, to get him to run faster. But no one was concerned with the condition of Mr. Supply. Even though Mr. Supply was increasing his pace, Mr. Demand was still running much quicker.

Uncle Joe was so happy that the first stimulus worked so well that he jolted Mr. Demand with a second energy drink. Now, Mr. Demand was sprinting at top speed. However, even though Mr. Supply was trying to run faster, obstacles on the trail slowed him down. He asked for people to work to help him run faster, but they were too distracted by Mr. Demand's stimulus, to aid Mr. Supply. Others declined to help due to the persistence of the pandemic. And some potential workers had left the trail altogether. He also needed a new pair of running shoes, but he couldn't get them due to a shortage of silicon.

So, no matter how hard Mr. Supply tried to catch up, he could not, and Mr. Demand was now miles ahead of him. This motivated Mr. Price to pedal faster than he had ever done since the 1980s. He needed to get ahead of Mr. Demand and slow him down so Mr. Supply could catch up. But he was pedaling so fast that his tires got overheated and are now highly inflated, and he fears a blowout, which could lead to a crash.

And so, it goes ...

Pent Up Demand

Whenever there is a product shortage, there is pent-up demand. This demand accumulates because it cannot be satisfied or fulfilled by the present supply. Product shortages, caused by various reasons, typically create pent-up demand for a product. However, if shortages persist, consumers may find substitute products, and some of the pent-up demand will dissipate.

It is challenging to measure pent-up demand, and many analysts and forecasters don't even try. During the Great Supply Chain Crisis of 2021-2022, I was able to estimate pent-up demand for Class 8 trucks by comparing more detailed information I had on pent-up demand for liquid tanker trailers.

It was an unconventional method, but it matched well with the number generated by our statistical demand model. I had the guts to present this estimate to a group of truck OEMs, and several told me privately that my pent-up demand number was very close to their internal estimates.

Pent-up demand complicates the forecasting process. How much of the order backlog is pent-up demand, and how much is due to panic ordering and placeholder items?

The accumulation of pent-up demand usually results in an increase in industry output to satisfy that demand. While pent-up demand is present, this formula represents monthly demand:

Total Demand = Standard Demand + Pent-Up Demand

When pent-up demand is substantial, production and sales will be robust for an extended time as companies strive to catch up. The conundrum here is that some pent-up demand is satisfied monthly once production (supply) increases. However, you don't know what the remaining pent-up demand is and what percentage of your monthly sales is regular demand versus pent-up demand. Your shipments can drop suddenly and significantly when that pent-up demand dries up. One segment of the commercial trailer market always experienced this cycle when recovering from an economic downturn.

How I explained it to people:

Imagine a large pile of rope. This is pent-up demand that has accumulated over time, and it is sitting stagnant. Now, you tie the rope to the back of the truck, and the truck moves forward, pulling the rope. You can only see the truck moving forward (the market is improving), but you can't see how much of the rope is left. At some point, there is no slack left in the rope, there is a loud snap, and the pile of rope is all gone – and so is all your pent-up demand.

And that snap can be demand suddenly falling below the forecast. This also can cause panic among Sales and The Cheese due to the unexpected sales dip.

Relevant Article

This was a blog post I wrote in April 2021.

Pent-Up Demand and a Roaring Twenties Redux

In '21, we see an economic recovery unlike ever before. Of course, I am referring to 1921, after WWI and the Spanish flu ended. But the country's mood now, as vaccines work to end this pandemic, is beginning to rise toward a euphoric state.

This, combined with tremendous government stimulus efforts, has caused the demand for goods to skyrocket. The GDP forecasts for 2021 continue to move higher. FTR forecasts 2021 GDP growth at 6.1%. In the latest Wall Street Journal survey of economists, the range is from 2.4 – 10.0%, with the average at 6.0%.

The ISM (Institute of Supply Management) Indexes, which are forward-looking, confirm there is robust demand present now and in the foreseeable future. The March Manufacturing PMI spiked almost four percentage points to 64.7, the highest reading in 37 years! IHS Markit's Index placed it at the second-highest reading ever. Likewise, the Services PMI jumped over eight percentage points to 63.7, an all-time high.

The economic shutdown in March-May 2020 created enormous pent-up demand in the economy. It produced a "sling-shot effect", where commercial activity was held back and then propelled forward rapidly. Therefore, substantial pent-up demand built up during the economic lockdown and was unleashed in the restart.

However, there was no pent-up supply, rather the opposite, in fact. Factories shut down, during well, the shutdown. Unfortunately, the restarts in many industries have been difficult. Manufacturers had to install COVID safety protocols. Workers have been reluctant to return to jobs either based on personal health concerns or generous government assistance. The global supply chain was also impacted, resulting in huge backlogs at the ports. Throw in February's polar vortex, and you get an unprecedented widespread shortage of components, parts, and industrial output.

The result is surging demand combined with pent-up demand, matched up against constricted supply. Of course, this creates more pent-up demand since manufacturing has still not caught up in the short term. Pent-up demand clouds the economic forecast because it is difficult to measure and determine how long it will take to catch up. The ISM numbers indicate pent-up demand is massive and still growing.

For an estimate of overall manufacturing pent-up demand, it appears that the current supply of Class 8 trucks is running about 20% behind demand. The shortage of semiconductors is impacting truck manufacturing, but many industries are not affected by this particular shortage. Therefore, let's estimate the total pent-up demand in all manufacturing at 15%. Meaning the supply is running 15% below total demand. Combine this with the enormous pent-up demand in the service industries due to the pandemic, and the economy is set up to surge in the next 12 months.

When there is another pandemic, perhaps world governments should expand the definition of essential workers to include those industries that should not be shut down because their products are essential to a restart. They could receive government loans to build inventory and quickly repay the loans when the inventory sells.

If you never understood the Roaring Twenties, you are about to get a very personal history lesson. But whenever I make that statement, the comment I always hear is: "Yes, but remember what followed the Roaring Twenties."

Well, advancements in medical intelligence and technology have enabled us to significantly limit the fatalities from COVID-19 versus the Spanish flu, on a population percentage basis. Let's hope our knowledge in the field of economics has made similar advancements.

Presentations

When giving formal presentations of your forecast, be careful not to throw a bunch of numbers on the screen. Instead, tell a story. People like hearing narratives and will remember them longer than charts and graphs.

Start by describing "Where are we now?" Then, "How did we get here?" Followed by "Where are we going?" Weave the charts and graphs throughout the story. Blend in your forecast assumptions as part of the story.

After finishing story time, you present the numbers. Do it confidently, but not arrogantly, and without those dreaded weasel words – might, maybe, could, etc. Then, list the assumptions and risks. Because these have already been mentioned in your story, you are repeating them for emphasis and review.

When someone challenges your numbers, you respond as always by discussing the assumptions. If there is disagreement over an assumption, conclude that discussion with: Yes, if you believe that industrial production will be weaker next year, then your forecast would be lower. If you do not stand your ground during the Q&A, some people will disregard your entire forecast.

Presentation Don'ts

1. Do not bore your audience by showing numbers, after numbers, infinitum.

2. Do not include too many slides – avoid PowerPoint Poisoning. The presentation masters can do 60 slides in 60 minutes – but you (and I) are not masters.

3. Don't make the presentation too complicated. The purpose is to clearly communicate the forecast and the supporting factors, not to impress people with your knowledge.

    Also, avoid the use of complex charts. If it takes more than three sentences to explain the chart, it may be too complex. This includes the dreaded "spaghetti" graph. If you use a spaghetti graph, just explain the key factor from the graph, not *how* the spaghetti was made.

Chapter 19

Demand Vs. Financial Forecasts

One of the most contentious issues you will encounter is the conflict over whether the demand forecast must match the financial forecast. Accounting pushback will be the toughest place for you to stand your ground. They have standards, and your process has standards. The goal is to try to work together for the good of the corporation.

Rule: THE DEMAND FORECAST DOES NOT HAVE TO MATCH THE FINANCIAL FORECAST!

Welcome To "Accounting World"

The insistence that the two separate forecasts must match comes from Accounting/Finance. In "Accounting World," all numbers need to add up perfectly. If the columns and rows don't match up, you have an error and must have done something wrong.

Accountants are obsessed with this concept because that is how things operate in the Accounting World. If the figures don't add up, they can have insane conniptions and become hostile.

The commotion can begin when Accounting plugs your forecast numbers into their financial forecast spreadsheet. The spreadsheet is then given to the Accounting Cheese, who considers the forecast unacceptably high. "We can't submit this to corporate," they howl. The Accounting Cheese then complains to The Big Cheese and anyone else who will listen that the forecast is wrong. Then comes the edict, not a request, to lower the demand forecast to more *attainable* levels and thus make the financial forecast more conservative.

I am not anti-accountant! Accountants perform an essential function necessary for your company's growth. They monitor spending, cost, and profit levels and ensure everything runs well. If these responsibilities were given to Marketing, your company would be toast in about three months.

The problem here with the demand forecasts is that Accounting is trying to force its standards on an issue outside of its domain. They are not being illogical; they are just overreaching.

The Issue of Forecasting Bias

As we have seen, all forecasts have some bias. Even computer-generated forecasts are biased since they assume that all historical trends will continue on indefinitely.

Demand/Production forecasts have a bias on the high side. It is usually beneficial to slightly over-forecast and have some resources unused rather than under-forecast and lose sales as a result. Lost sales typically occur if the supply chain is tight, labor is constrained, and the company lacks "flex capacity".

The cost of lost sales from insufficient inventory or restricted production is almost always greater than excess inventory and material costs. In addition, the cost of lost sales may extend into the future if customers are upset with your stockouts. They may try your competitors' products and permanently switch. You may also miss the opportunity to acquire new customers if they can't get your product. This is behind the whole concept of "safety stock".

Of course, there are inventory-carrying costs that should not be dismissed. Typically, these become problems when they build up over time. Usually, an over-forecast results in a reasonable rise in the inventory of parts/materials. Part of inventory management is dealing with monthly fluctuations while minimizing the safety stock. It is much easier for Manufacturing to deal with an over-forecast and subsequent scale-back than an under-forecast and a ramp-up.

Also, The Cheese is more displeased with an under-forecast than an over-forecast. It is never good when The Cheese becomes aware that your forecast was out of standard. If the under-forecast results in lower sales, Sales will loudly complain to The Cheese, often inflating the impact. It usually takes a few months of over-forecasts for Manufacturing to notify The Cheese due to excess parts inventory and lower plant efficiencies. In addition, over-forecasts are usually more easily explained (have those forecast assumptions ready!) than under-forecasts (Whoops, I didn't know about those big orders).

Therefore, although the goal is to get as close to the actual number as possible, production forecasts have an inherent upward bias. Even though I never measured it, my production forecasts were probably higher than actual at least 60% of the time. That's fine because my production planners were well aware of my biases due to the Forecast Accuracy Report, and they adjusted their production plans accordingly. Remember, your job is to provide the most accurate forecast possible. Their job is to take that forecast and develop the best production plan possible. That being said, because you are human, you have biases.

Conversely, financial forecasts have a significant bias on the low side. It is called "The Price Is Right" forecasting; it comes as close as possible without going over. There is a substantial psychological factor with financial forecasts. Say Mega Corporation has a terrific quarter, with profits up 20% from a year ago. But sadly, the profits missed the forecast by 1%, and the stock sank by 2% that day due to the headlines. Your forecast accuracy was superb, but you were the loser that day based on The Price (or the Forecast) Is Right.

If sales/profits exceed the financial forecast, everyone is happy and walks around with giant, goofy grins. The company has done great! Heck, The Cheese may buy everyone some shrimp! The company has done tremendously and is thriving. If the sales equal the forecast, the company has merely met expectations—yawn, meh.

But woe to the company that consistently or significantly misses the sales forecast. All corporate hell breaks loose. Corporate, The Board, or Ownership may threaten The Cheese's job. This results in a cascading torrent of screaming, threatening, and gnashing of teeth. And the political blame game commences in full force. Unfortunately, this can reach all the way down to the lowly forecaster. "We need to blame someone for this! Hey, why was the forecast so high?"

Even though the competition ate your lunch that quarter, your suppliers didn't deliver on time, and your incompetent sales team bungled some big potential deals, that target on you is flashing bright red. Document your assumptions on all financial forecasts and note somewhere who signed off on the final forecast. Wretchedly, The Cheese who approved your forecast may be the same person who calls for your head.

The forecaster has some power! Not the power to change company performance, but the power of performance perception. And, of course, that power, if not managed, can cost you your job. When push comes to shove, you may get pushed out the door even if you did nothing wrong.

There are no good reasons to be optimistic or aggressive with an actual financial forecast (we will dismiss flip comments made in the media to boost investor confidence or raise venture capital). It is advisable to keep your financial forecasts conservative but reasonable. Unfortunately, the factors discussed previously can tempt companies to "sandbag" or "lowball" the financial forecast. Submit a low forecast – greatly exceed the forecast – bonuses!!!!!!!!!

Merriam-Webster explains:

"By the mid-20th century, sandbag was being used by poker players to describe the act of pretending a strong hand is actually weak in order to draw other players into raising the bet. This use of sandbag has since evolved to refer to a general strategy of misrepresenting one's intentions or abilities in order to gain some sort of advantage."

Most of The Upper-Cheese know The Lower-Cheese will attempt to sandbag forecasts, so this strategy seldom works. And when it does, it is usually temporary. If the company exceeds the forecast by 30%, it is obvious some sandbagging has occurred, leading to increased scrutiny on the next forecast.

Sales also benefits from a conservative financial/sales forecast. Sales may strongly lobby for a lower forecast when sales quotas and bonuses are based on those numbers.

Different Factors in Different Forecasts

Another factor to consider is that demand forecasts are focused on total orders requested to be shipped in a given month. These orders are subject to cancellations and rescheduling (move-outs). The demand forecast may also include potential orders that sales have yet to close, but Manufacturing must be prepared to build. Financial forecasts are focused on what is shipped/billed. You will invariably run into a situation where a lot of finished product sits near the dock on the last day of the month and gets shipped the following month. This displays the difference in the forecasts. Demand forecasts are what you expect to need to build, and financial forecasts are what you expect to ship.

Resolving The Conflict

To Repeat: THE DEMAND FORECAST DOES NOT HAVE TO MATCH THE FINANCIAL FORECAST!

The challenge is reconciling the demand forecast with an upward bias to the financial forecast with a downward bias. The simple solution would be to average the two. While this might eliminate the biases, the two forecasts exist for a reason and serve both functional and company political purposes.

But rest assured, Accounting will want, maybe demand even, the forecasts match for the reasons previously explained. This issue arose two times when I was forecasting at a previous employer (how I handled it will be described later). The sham is that when Accounting says the forecasts have to match, they really want the demand forecast to match their financial forecast. They surely don't want the financial forecast increased to the demand forecast, do they? And you will never suggest that approach because if you do, the financial forecast will be too high, putting your job in jeopardy.

When Accounting asks that the forecasts match, you can ask them why they must. Of course, this is akin to challenging a sacred tenet of a religion. "Well, they have to!" is probably the response you will get. In reality, Accounting shouldn't really care at all about what is in the demand forecast. It doesn't impact them a bit unless a series of high forecasts have pushed inventory/material carrying costs up to unreasonable levels.

Your response to this request should always be:

"The demand and financial forecasts are two different forecasts used for two distinct purposes. The forecasts need to be based on similar assumptions, but they should not be identical."

If that doesn't work, there is a summary at the end of the chapter that you may copy and give to Accounting. I was initially going to tell you to copy this chapter, but after all the swell things I said about accountants, that is probably a bad idea. If you are an accountant reading this – I still love you – but please don't require the two forecasts to match.

Related But Not the Same

If the forecasts don't have to match, we can just do whatever we want, right? No, you can't. The demand and financial forecasts are not twins, but they are cousins. They need to be related to each other. My general rule is that the units forecasted in the demand forecast should be at most 10% higher than the units used to generate the financial forecast. If you are doing both forecasts, those are guideposts. If two people do them, they must work together under this guideline. It is critically important that the product mix in the demand forecast is reflected in the financial forecast. If not, you risk overstating profits, which, of course, can be worse than overstating sales.

There should be a list of assumptions for the financial forecasts, and these should differ slightly, but notably, from the demand forecast assumptions.

For example, let's say the demand forecast assumes a market share gain of 2% for Product A for the forecast period. The financial forecast may assume only a 1% gain. This is entirely acceptable. These are predictions about the future. The reality is you do not know what the market share gain will be. You are making your best deduction (never a guess!) based on what you know. It may turn out that there is no market share gain at all. Heck, Product A might even lose some market share. (Sales & Marketing, you got some *splainin'* to do!).

If Forced to Match

Accounting may put the screws to you and force the demand and financial forecasts to match, resulting in you reducing your demand forecast. If this happens, first appeal to your boss. If that doesn't work, you have lost the "control" element of your forecast. Lower your forecast and concede defeat.

Of course, a grand compromise would be to average the two forecasts. But this brings up the old line: "I would agree with you, but then we would both be wrong." Any action that raises that financial forecast puts you in political danger – "Yeah, our financial forecast was too high because the forecasting person averaged the forecasts!"

But you can't let the silly-ass corporate games interfere with your responsibility to work with Production to provide excellent customer service and cost efficiency, can you? So, go to your contact person in Production/Materials and explain that you had to lower your initial demand forecast to match the financial forecast, and the forecast you are sending them is about 10% (or whatever) too low. Warn them to take this into account when preparing their production plan.

At this point, you want the production planner to ask:

Can you send me your original version?

Offering the original version has political pitfalls, but merely providing it to Manufacturing after they ask for it allows for sufficient political cover. Just don't name the file a "forecast." "April alternative" should do the trick. If asked later about this, you can say you provided the data upon request. If your contact doesn't request the original forecast, you still have alerted them that the official demand forecast has been artificially lowered.

You can measure your accuracy against your initial forecast if you choose. It's your measurement, so you do have control over that. There is little risk of anyone noticing what you measured against, and if someone does, then explain that you wanted to measure against your forecast before someone else changed it.

Why the Demand Forecast Should Not Match the Financial Forecast

1. The demand and financial forecasts are two different forecasts, developed for two different purposes, to be used by two different company departments.

2. The demand forecast is focused on potentially what needs to be built that month. The financial forecast is focused on what is expected to be shipped and billed that month.

3. Manufacturing uses the demand forecast to plan production. Executives use the financial forecast to understand and communicate the corporation's/business unit's current business conditions. Accounting/Finance uses it to secure corporate financing, plan resources, and anticipate profit performance.

4. The demand forecast has an upward bias. If the forecast is too low, lost sales can result, and factory efficiency suffers. The financial forecast has a downward bias. If the financial forecast is too high, management/stakeholders are unhappy, and people can be held accountable for "missing" the forecast. It can also negatively impact cash/resource management.

5. The demand forecast is based on potential orders, which may or may not materialize. The financial forecast is based on shipments/billings, which are subject to product/supply chain delays, order cancellations, and shipping delays.

6. For these reasons, the demand forecast and the supply forecast should not match. However, these forecasts are related, and there should be a manageable gap between them. A good general rule is that the forecasts are within 10% of each other. Take care to make sure the product mix is consistent with both forecasts. If it isn't, profits can get overstated/understated.

Relevant Story #1

Soon after I started distributing my monthly demand forecasts, the Accounting Cheese demanded that my forecast match the financial forecast. I had found an old journal article that addressed the subject and presented him with a copy. And that worked—for a while. The scenario was repeated two years later, and the article worked again! So, copy that section above and "fire at will" when the subject is raised.

But I didn't totally get off the hook. Because the forecasts didn't match, Accounting requested that I also do the financial forecast, which I did. This will be discussed in a subsequent chapter.

One Final Comment

As previously stated, I did not want the industry forecast at my final employer to be biased, but in this section, I write that demand forecasts inherently have upward biases. This is not a contradiction.

Company demand forecasts used for production planning have upward biases. The demand forecast at my final employer was for the entire industry. I wanted that forecast to be bias-free because our customers would inject their own bias and knowledge into their specific company demand forecasts. If they thought our forecast was too aggressive, they would adjust downward and vice versa. So, we needed to provide them with the most accurate baseline forecast and best assumptions possible.

On occasion, the forecast at my final employer contained what I call "competitive bias." There were competitors and financial analysts who forecasted the same markets. I usually didn't care what their forecasts were, but if the gap between our forecast and the competition got too wide, some customers would get nervous, and my phone would ring.

During these discussions, I would learn where we were relative to the competition's forecast. If it was an outlier to the competition's forecast, either high or low, in the next forecasting meeting, I might resist moving our forecast if it increased the gap. We already owned that territory in the forecasting map; there was no need to pay a higher price if our assumptions were incorrect. Fortunately, we were seldom wrong.

Oh yes, and remember: **THE DEMAND FORECAST DOES NOT HAVE TO MATCH THE FINANCIAL FORECAST!**

Chapter 20
How To Estimate Anything

Often, you are asked to estimate or forecast impossible, impractical, or downright silly things. Of course, many of these requests come from THE CHEESE. The best reply is, "That number would be hard to get to. There isn't enough reliable data available to do that." This is not as dismissive as the "I don't know" and, in most cases, ends the discussion.

There are instances where the request is legitimate but challenging to fulfill. Or where THE CHEESE *expects* an actual answer to their unrealistic request.

Therefore, if you must provide an estimate on something obscure, follow these steps:

Step 1 – Establish your parameters

Set your bottom parameter as one, or close to one, since you know that your factor does exist. The top parameter is often the total possible population of whatever you are trying to estimate. For example, if the total possible population of your factor is a million, your initial parameter range is somewhere between one and a million.

Step 2 – Start tightening the range

Gather all relevant information or even opinions that enable you to add to the bottom or subtract from the top of your range. To accomplish this, you must make several assumptions (document them), sometimes broad ones. This is necessary because of the task's difficulty and complexity.

Step 3 – Continue to narrow the range

As you work on the answer, you will generate new questions and, thus, new sources of information, enabling you to narrow the range further. Of course, almost everything in this process will be based on more assumptions.

Step 4 – Finalize your range

After narrowing the range as much as possible, review your assumptions based on how certain you are about each assumption and then make adjustments as necessary. For example, if your narrowed range in the example is 200,000 to 400,000 but five of your most positive assumptions are iffy, narrow the range to 200,000 to 300,000.

Step 5 – Apply the smell test

Does your number look reasonable? How does it compare to other parameters that you do know? For example, if your estimate of skyscrapers in Helena, Montana, equals 20% of the known total in New York City, your estimate needs to be lowered, and you have more work to do. If your estimate makes sense, you are finished.

You will be asked how you came up with the estimate since whoever asked knows the task is complex. Refrain from giving a detailed description of the process, or you may find yourself in an extended discussion or debate. Something like, "I started with a base number of X, and then made various assumptions about Y to get to this range."

An Example

Let's say we want to estimate the number of people in the United States with freckles on their faces. Yes, it is a whimsical task, but not much stranger than some requests you will receive.

The U.S. population in 2021 was 332 million. That is your high number. And 10 can serve as the low number, an estimation of the people you know who have freckles.

We know many red-haired people have freckles. Estimates from the Internet say that 1-2% of the population has red hair (even this number is not that exact). Let's use 1.7%; this gets us down to a high number of 5.64 million. Another Internet estimate says 80% of redheads have freckles. This narrows the number to 4.51 million.

The 4.5 million is a decent estimate for redheads. But some blondes, as do a small number of people of other hair colors, also have freckles. Interestingly, only 2% of the population is naturally blonde, giving us a blonde number of 6.6 million. There is no estimate on blondes with freckles. So, let's assume that 50% of natural blondes (because they have lighter skin) have freckles. This gives us another 3.3 million people.

The total of blondes and redheads with freckles gets us to 7.8 million people. Let's add in a half-million other hair types with freckles for a total of 8.3 million. If we use 1.5% for the redhead population, 25% for natural blondes with freckles, and a half-million other hair types with freckles, the estimate drops to 6.1 million. So, a reasonable range would be 6.5 to 8.5 million.

Comparing the 8 million estimate to the general population yields a U.S. freckle percentage of 2.4%. This passes the smell test. I would have estimated that the number was higher than 1% and below 5%.

Therefore, the estimate is 8.3 million with the following assumptions:

- 1.7% of the population are redheads
- 80% of redheads have freckles
- 2% of the population are natural blondes
- 50% of natural blondes have freckles
- 500,000 of other hair colors have freckles
- All the freckle estimates assume freckles on the face.

Trust the Process

This process is effective and efficient in developing reliable estimates. It keeps you from saying the dreaded "I don't know" or "I can't do that". It can win you status points with your boss, and The Cheese. Usually, your number will not be questioned because of the task's difficulty. It is typically not worth arguing over. If The CHEESE insists their SWAG number is better, go back and change your assumption, and voila! – The CHEESE were right after all. And we all know how much The CHEESE like to be right!

Relevant Article

I do believe I can estimate anything, especially with all the information available on the Internet. Here is an example of the process from my humor book "Turkey Terror At My Door!" The essay is meant to be humorous, not scholarly, but it illustrates the abovementioned process.

Counting All Angels – That was One Big Hark!

*"The world in solemn stillness lay
To hear the angels sing…"*

So many of our favorite Christmas carols refer to the angels singing and announcing the birth of Christ. The basis for this is Luke 2:13-14:

And suddenly, there was with the angel a multitude of the heavenly host praising God and saying,

"Glory to God in the highest, and on earth, peace, good will toward men."

This event was so magnanimous and spectacular that a "multitude", "great company", or "legion" (depending on the translation) of angels were dispatched to Earth to sing praise to God. But just how many angels were there in this choir?

Now, Don, why would you even ask that question? No one knows how many angels there were, and it's impossible even to estimate!

But these are the types of questions my hyperactive mind generates. And once the question enters my brain, I must answer it.

Furthermore, it's what I do for a living. I forecast things that are difficult to forecast, and I make estimates about things that are hard to figure out. So, I am going to try to do this seemingly impossible task.

And I assure you that when my coworkers read this, they will laugh out loud because they have to deal with my lunacy about stuff like this all the time.

The question is: How many angels appeared to the shepherds, singing praises to God, at Christ's birth?

Surprisingly, the answer to this question may be zero. You see, Luke 2:13-14 doesn't actually say the angels sang—it says they were "praising". However, we have always assumed the angels were singing those praises because we praise God through song in our culture. So, all the Christmas carols reference singing angels.

Now, if the angels weren't singing, they were at least chanting. It had to be scripted praise for the large group/choir of angels to be understood by the shepherds. I know this because years ago, I had to drive through a group of striking workers at our factory who all yelled different insults as my car passed by. It was comical because it all sounded like static, and I couldn't hear any actual words. And that's the literal purpose of cheerleaders, right? - To coordinate the cheers so the players get the message.

Therefore, those angels had to be singing or chanting in unison. And while some Gregorian monks would argue for the latter, I think we can assume the angels were singing if we interpret the word "saying" in Luke as "communicating". And such a large choir would have had to practice these praises in advance.

To estimate something difficult, I first determine a higher number that the estimate can't exceed and a base number that the estimate should exceed. Then, I make assumptions to define a number somewhere in between.

The High Number

Scholars debate about how many angels exist in total. The Bible implies that the number is enormous. If we interpret the term "myriad" of angels literally (Revelation 5:11), there are at least 100 million (100,000,000) angels (10,000 times 10,000). It is unlikely that God would have sent all 100 million to Earth simultaneously. Now, if it were me, I would be tempted to "send the house" for my son's birthday, but I'm not God, and everyone should thank God for this.

However, the upper limit to our question is **100,000,000** angels.

The Low Number

The record number of people in an earthly choir (defined as singing the same song in one place) is 121,440, set in India in 2011. Let's assume that our angelic choir has to be greater than anything that can be accomplished on Earth. Considering future efforts to break this record, I will set the minimum at **200,000** angels.

The Range

Therefore, at this point in the analysis, the range is between 200,000 and 100,000,000.

The Assumptions

To narrow the range, I have to make some assumptions. These assumptions are complex because they involve God, angels, and an incredible event that occurred long ago.

Assumption #1

God did not send all the angels to Earth because the shepherds could not have seen them all. This is an excess of angels. Yes, this was a big event (again, it's His son's birthday!), but all the angels were not needed to accomplish the task. We know from the Bible that there are different types and ranks of angels, so let's say the trip to Bethlehem was a reward for the top 1% of the angels. Yes, just like a Mary Kay bonus. That reduces the number down to **1,000,000** angels.

Assumption #2

Could one million angels hanging up in the sky be visible to shepherds on the ground? The number of angels needed to completely fill the sight line of shepherds on the ground can be calculated using equations containing things called "parabolas"—but I'm not a mathematician, so I won't attempt this. You would have to make assumptions about the size of an angel, but I guess it took fewer than 1 million angels (1,000 rows of 1,000) to fill the sky.

Assumption #3

The decibel level of the singing angels can't be high enough to wake the people in Bethlehem. The announcement of Christ's birth was only intended for the shepherds. Supposedly, this is because they were the lowest in that society, and Jesus came to Earth for all men. Hey, the only job requirement for a shepherd is that you must be more intelligent than a sheep. Oh, and you're working the third shift. Yes, you are on the bottom rung of that city, and yet you're worthy of hearing this good news first.

It may have been a silent night up to that point, but hundreds of thousands of singing angels make some noise. The praises didn't alert the townspeople; more importantly, it didn't wake the baby.

Let's assume the angels sang softly and were positioned facing away from the town to reduce the noise factor. But again, the more angels, the more decibels. This also argues for a total number of under one million.

The Call:

The assumptions get the total to between 200,000 and 1,000,000. The space and volume assumptions push the total under 1 million. So, I will put my estimate at approximately **500,000** angels. As it is estimated, let it be said.

So, every time you hear, sing, or even chant a Christmas carol this season that references singing angels, think about just what a half-million angels might look—and sound—like. It's probably better than the best Dolby system ever created by man!

CHAPTER 21

Dealing With Caustic Company Politics – And Other Corporate Crap

Every organization suffers from internal politics. The issue exists in large corporations, small companies, non-profits, government entities, and maybe even your local bridge club. This occurs because you assemble a group of people with diverse personalities, expectations, and goals and expect them to work together as a highly productive, unselfish team. Unfortunately, in most cases, they compete against each other for raises, promotions, status, psychological needs, personal relationships, and other limited resources.

It's as if you locked some rats in a box (the office) and threw a limited amount of food to them every day. Corporate politics is dog-eat-dog. But that is an insult to dogs. It is more like rat-eat-rat.

From my experience, the larger the organization, the worse the political environment. Of course, this varies considerably depending on the company. In many cases, this depends a lot on The Cheese. At one of my previous employers, THE BIG CHEESE actually encouraged his rats to fight it out. He believed it maximized company performance and enjoyed watching the battle, similar to a Roman emperor.

I wrote this book to help you survive, flourish, and stand your ground as someone with a vulnerable job in a caustic environment. If you believe I am too negative and am overstating the toxicity of company politics, well, yes, I am jaded. And here's why:

Relevant Story #1

Dealing with evil sabotage ...

As mentioned previously, I was fighting for my corporate life at a previous employer. I was trying to do three different jobs, and the division's rapid growth had outstripped my ability to do any of them well. My pricing system needed to be reprogrammed immediately, and corporate IT didn't prioritize my fix. Smelling blood, one of my coworkers deliberately tried to get me fired. She blatantly lied about me to various people throughout the company. And she delighted in pointing out any pricing errors to management and The Cheese. I have no idea what caused this awful behavior.

It got this bad. A customer service rep had to quote a price to the customer for a future order right before an upcoming price change. I hand-calculated the new price, and we got the order.

I entered the new prices into the computer system, being very careful due to the recent pricing errors and my perilous position. I printed a hard copy of all the price changes as an additional check. I told the customer service rep he could now enter the order and the system would price it correctly.

A half-hour later, the customer service rep sprung into my office and announced that the system price differed from what I had given him. I compared the prices in the computer with my paper printout and found that someone had gone into the system and changed the component prices I had entered just an hour earlier. I corrected the prices, and the order was entered. Of course, I printed another hard copy for backup.

Only two other people had access to the pricing file, and one was an IT administrator who had never accessed the file before. Therefore, it was easy to determine that my nemesis had been sabotaging my price files, maybe for an extended time. Even though I had evidence, I couldn't accuse her of wrongdoing for several political reasons. But showing the printouts to my boss saved my job.

Unbelievably, many years later when we both had left the company, that vile person actually invited me to connect on LinkedIn. That was a hard "No". Yes, unfortunately, these are some of the people you have to deal with in your career.

A Perilous Job in A Hostile Environment

As previously described, being a forecaster is akin to being a 10-point buck on the first day of hunting season. You are a large and attractive target. The essence of corporate politics is the blame game, and everyone will try to shift the blame to the forecast whenever possible.

Some people you work with are nasty, vengeful, mentally ill individuals. They will view the forecaster or business analyst as an easy target for bullying. In the real world, if you stand up to the bullies and know how to push back appropriately, the bullying ceases. That is one of the key benefits of this book – being able to stand your ground in tough situations.

Playing Politics

I am not an expert on corporate politics. In fact, I was lousy at it. I didn't schmooze my bosses or The Cheese enough. I didn't tow the company line or blindly follow ridiculous orders. I was too smart for my own corporate good. I naively believed my superior work performance was adequate to achieve raises and promotions (please stop laughing now). While this philosophy maintained my integrity, it undoubtedly cost me jobs and promotions.

Let's focus on your survival as a business forecaster/analyst in the corporate jungle. You have enough to contend with in corporate politics without the burden of dodging those forecasting potshots. But corporate politics always leads to conflict. Here are some strategies to reduce the conflicts.

The Political Players

Your Boss

You need your boss in your corner and protecting your back or maybe even your backside. When push comes to shove, and it will, you need them to deflect the criticism and even push back if necessary. Without their support, it's difficult to exist, let alone flourish, as a forecaster.

Therefore, your boss needs to know (not necessarily fully understand) your forecasting process. They need to believe you are committed to continuous improvement and tightening forecasting accuracy. Even during "forecasting slumps", if your boss knows you have a sound process and are committed to doing good work, they can defend you when attacked by whomever, including The Cheese.

For these reasons, frequent communication with your boss is essential. If there is a forecasting issue, your boss needs to hear it from you rather than being blindsided in the staff meeting in front of The Cheese. It is part of the "no surprises" strategy described earlier. If your forecast was not accurate, make sure the boss knows this and the reasons why. Also, notify them of conflicts with other departments.

Even though your boss is getting a copy of the Forecasting Accuracy Report, they probably ignore it, especially if you are doing a good job. When you have an excellent forecasting month, tell them, "Hey, did you see I really nailed that March forecast?" This will remind them of the report and enhance your status at the same time.

Accounting

A significant flashpoint of conflict for forecasters is the Accounting department. The issue here usually involves the financial forecast. Accounting may use your demand forecast as the basis of the financial forecast, or just match it (not a great idea, as previously detailed).

While the Big Cheese will rarely review the demand forecast, they will examine the financial forecast in great detail. The financial forecast may be reported to stockholders and stakeholders and debated on CNBC.

There is corporate hell to pay when the financial forecast misses the mark. People are upset at The Big Cheese, who excoriates the Next Cheese, who rips into the Next Cheese, until the Accounting Cheese, who, of course, may place the blame on the lowly demand forecaster.

Accounting may yell loud and long at you because they yell at everyone in the company. They consider berating part of their job, and some secretly enjoy it. But Accounting is not the enemy; they should be regarded as part of your forecasting team. Communication is, again, the key to cooperation.

If the financial forecast is related to your demand forecast, then your accounting liaison should be aware of your forecasting process and review your assumptions monthly. This person can attend the Forecasting Meeting if it helps. Your assumptions should be incorporated to support the production forecast. If Accounting disagrees with your assumptions, they can adjust *their* assumptions and *their* financial forecast accordingly.

Suppose the financial forecast for a given month is 7% below the demand forecast, but the actual financials come in 10% below forecast. In that case, that usually means you also significantly missed the demand forecast. That's why the Forecast Accuracy Report is essential. If Accounting already has the information on which assumptions were incorrect, it allows them to give credible answers to The Cheese and stop the chain of verbal torching before it gets to you.

Relevant Story #2

Sometimes compromise is the best option ...

When I refused to make the demand forecast match the financial forecast, the second time the issue arose, the Accounting Cheese threw a hissy fit. "Well, if you are not going to make them match – then YOU can do the financial forecast, too!"

I really didn't want this added responsibility, but I wasn't going to make the demand forecast match the financials. So, after consulting with my boss, I agreed to do the financial forecast to mollify the Accounting Cheese.

Of course, there was considerable risk. I was now vulnerable to taking heat over two forecasts, and the financial forecast was more visible to The Cheese. However, I took on the task of setting the financial forecast 5-10% below the demand forecast each month, as recommended previously. I tracked the sales accuracy of the forecasts, and the new arrangement worked well. Note, I don't recommend publishing and distributing the forecast accuracy of financial forecasts due to the corporate political climate. On bad months, this could produce weeping and gnashing of teeth.

While the financial forecast was extra work with extra risk, there was an unanticipated benefit. The conflict with Accounting over the forecasts virtually disappeared. As I explained to people then, "How can I be in conflict with myself?" I could align the financial forecast exactly how I wanted it with the demand forecast. The forecasts were related but different, and everyone was happy.

Preparing the financial forecast also insulated me from some criticism. If confronted by The Big Cheese, the Accounting Cheese was reluctant to blame me for an inaccurate financial forecast. If he did, The Big Cheese could rightly ask, "Why is Don doing *your* forecast?"

Also, if the Accounting Cheese complained to me about the financial forecast, he knew I could refuse to continue doing *his* work, which was benefiting *his* department. So, we had a stable détente situation with minimal conflict.

Manufacturing/Materials

Your process is designed to create a "team forecasting" approach where you and Manufacturing work closely together to achieve maximum efficiency and effectiveness. A suboptimal forecasting process can generate considerable departmental friction. Forecast discussions can degrade into shouting matches like those first forecast meetings at a previous employer described before. Everyone is under pressure to perform; throw some toxic politics into the mix, and things can quickly explode.

If your company experiences significant conflict over the forecast, please use this book as a reboot. Explain to your manufacturing cohorts that you want to revise your process based on this book and ask for their help. If they are reasonable, you are on your way to "team forecasting." If they reject your proposal, continue to revise your process, but watch your front, back, and sides while doing it.

Despite doing everything this book proposes, you will never win everyone over. The Plant Managers are usually the harshest critics. They have demanding jobs and are often expected to perform perfectly. Therefore, they expect the forecast to be perfect. When the forecast is outside the accuracy standards, it impacts them the most, and they scream the loudest.

If you have remote plants, I would suggest you visit the plant or meet the plant manager when they visit the home office so they put a name to the face and understand there is a professional forecaster in the house, working hard to provide reliable (never perfect) forecasts. I do not recommend including plant managers in the Forecast Meetings because of their high expectations. I knew we had a strong team forecasting approach at a previous employer when a plant manager began to rip into the forecasts during a conference call, and the Manufacturing Cheese shut him down swiftly and forcefully.

The Cheese

There is a delicate balance to how visible you want to be to The Cheese. As previously stated, when the forecast catches the attention of The Cheese, it's terrible for you. Someone has complained and tipped them off to a problem, not only criticizing the forecast but criticizing you.

That's why your boss needs to know when and why you missed the target so they can blunt the criticism upfront. You need to know your incorrect assumptions when The Cheese unexpectedly shows up at your desk with *that* stern expression on their face, as they are prone to do.

If The Cheese confronts you over the forecast, stay focused on those assumptions. The Cheese may have better insight into those assumptions based on their contacts and industry conversations. If appropriate, offer them a copy of the Forecast Accuracy Report to show you are in control of the process because The Cheese like employees who have control of their responsibilities.

If The Cheese is just torching you because someone complained and they feel the need to chastise your job performance, there is not much you can do. You can discuss assumptions and what improvements to the process you can make in the future. DO NOT give them a copy of The Forecast Accuracy Report if the discussion goes off the rails. They will just focus on the red ink in the report and wonder if you should remain at the company.

How visible you want to be to The Cheese depends on your company, personality, and forecasting performance. I don't recommend the hermit strategy because you will only get noticed when your forecast is off. A low-profile approach can help you avoid corporate downsizing if you and your forecasts are accurate and thus go unnoticed. It also depends on the disposition of The Cheese. Some are hotheads, sheepheads, or both, so act accordingly.

A good strategy is to raise your visibility as your forecast accuracy improves. Regardless, you are still a forecaster with a target on you, so there are limits, especially in a hostile corporate environment. Stand your ground – until it makes career sense to retreat.

Playing Defense

A forecaster in a perilous political setting must always be playing defense. Again, you are like a hockey goalie. Various people in your company, from all different departments, will fire shots at you from all directions. Heck, even your boss, if you work for a butthead, may even deliver a slapshot.

Your goal is to block every shot. I have provided the tools (think hockey equipment) to do that in this book. Unfortunately, you cannot let those shots go unimpeded. You must be like a national politician and respond to every criticism you get because they will stick to you if you don't. If too many stick, you won't last long in your job.

If you fail to respond to the criticisms, you risk becoming a corporate punching bag. Unfortunately, some of your coworkers are mean, vindictive people (as in Relevant Story #1) who greatly enjoy tearing people down. You may even work with a few psychopaths (I could create a Top 10 list from my career). But once they see you can effectively block the punches, most will stop punching because it is no longer fun.

The more significant challenge is to defend your forecast and block these attacks with as much professionalism, tact, and calmness as possible. In the tough corporate environments, concentrate on defense. You can't really go on offense because of the risk. If you are perceived as arrogant, you invite snide remarks the next time you blow a forecast. Stand your ground – but don't go on the attack!

Additional Reminders

Here are a few reminders about navigating the dangers of corporate political jungles.

The Forecast is Never Wrong

Remember: The forecast is never "wrong". It was outside the accuracy standards due to some invalid assumptions.

Your standard response to "The forecast was wrong!" should be something like:

"The forecast missed the accuracy target last month due to some incorrect assumptions about the Q2 market share, which proved overly optimistic."

You Never Guess

Remember: You never guess. You deduce based on the best available information that you have. So:

"You really had to just guess at it, right?"

or

"Nobody knows, so it's just your best guess."

And your reply:

"Oh no, I never guess. I deduce on the best information I have at the time."

Don't Apologize for The Forecast (Unless you have to)

My policy was never to apologize for a forecast, even when it was way off, unless I overlooked something obvious or made a math error. Instead, I would explain why my assumptions were faulty, admitting to the mistakes but not apologizing for the forecast. I know many people will disagree with this, but apologizing in this case is a sign of weakness. A forecaster cannot show weakness because of all those company sharks. That's why you should stand your ground when challenged.

I did have a boss who would keep pounding me until I admitted my mistake and apologized. He was very forgiving of even big mistakes if you apologized. But he required an apology. This created issues for me in cases where I believed I had done nothing wrong and thus had nothing to apologize for. Those conversations tended to be long and unpleasant. So, there were a few times I ended up apologizing for mistakes I didn't make.

Of course, admitting when you are wrong (but not apologizing) about assumptions or making bad calls is always acceptable and beneficial. It builds credibility and shows confidence, unlike those politicians who will not own up to any errors that are obvious to everyone. At conferences hosted by my final employer, I would show our forecast accuracy statistics from the past year, even when the performance could be considered substandard. It enhanced my reputation in the industry and showed I was a competent, credible forecaster in control of the situation.

It's Your Job – So Fight to Keep It

Corporate politics are dangerous for all jobs but much more perilous for forecasters. So, block every shot that comes your way. Never lie, but use every tool at your disposal, every trick in your book, to survive the corporate insanity and stand your ground.

When You Stand Alone – But Your Numbers Are Right

We will end this chapter with stories illustrating the challenges of being a forecaster, business predictor, or marketing researcher in a highly politicized environment. Sometimes, you must stick to your numbers even when others doubt them.

Relevant Story #3

Don't doubt good information ...

We had launched a highly successful new product at a former employer. The product used new technology to improve a medical procedure for both patient and doctor. A central selling point was that it eliminated the patient's pain in what could be an agonizing experience.

Nine months after the introduction, I was tasked with conducting a marketing research study to collect information from doctors about their experience using the device. The survey had a high response rate and collected much useful information.

A surprising finding was that four respondents had written in the "Any Other Comments" section that their patients had experienced minor pain during the procedure. I included this information in the "Additional Information" at the end of the survey results report. I distributed a draft to the marketing team for review before issuing the final version throughout the company.

Soon afterward, the product manager and her assistant burst into my office, and they were livid. I have never upset anyone more with a report in my entire career.

"This product does not cause any pain, period!" she shrilled. "You can't put that in the report because it is so false! They must be doing something wrong!" After demanding, not requesting, that statement be removed from the final report, they scurried to my boss's office, making the same loud, impassioned demand.

When they left, my boss sauntered down to my office with a worried expression. "They're furious," he bemoaned. "They want that comment stricken from the record. What do you think we should do?"

I argued the information needed to be included because that is what the results indicated, but I would rephrase the comment. I didn't know for a fact that our device was causing pain, but I did know that four respondents said it did, so that fact went into the final report. The product managers were not pleased, but it was an indisputable fact.

Several months later, we contracted with a doctor to travel around the country, giving seminars on using our device and promoting the benefits. Before he started the tour, we brought him into the office to hear his prospective presentation. So, we are all gathered in the conference room, and then the doctor says this:

"Right after I prep the patient and just before starting the procedure, I tell the patient that they are going to experience some minor pain at the beginning."

Of course, pandemonium broke out as the product managers freaked out again, demanding an explanation. I looked at my boss, who was already staring at me bugged-eyed; his jaw dropped. At that moment, I wanted to jump up on the conference room table and do an elaborate touchdown dance. I would have spiked my calculator if it wasn't breakable.

Relevant Story #4

The definition of corporate insanity ...

A previous employer had decided to introduce a major new product. These medical-use composites were hot growth products, and the market was dominated by well-entrenched offerings from the "big boys". My company also wanted to be a big boy in this growing sector and partnered with a European manufacturer to sell their "new, improved" composite in North America. My company was sure the new composite could grab lots of market share away from the established players and had already hired two new people with composite experience to champion the marketing push.

I was tasked to conduct a market research project to determine just how much market share we would snatch away from the competition and estimate initial sales to provide a forecast to the producer. Of course, it would have been better to do the study before those people were hired, but The Big Cheese was so sure we had an absolute winner.

I created a product concept test (my master's thesis subject). I listed all the expected product advantages, avoiding over-hyping it. The Cheeses approved the concept statement and questionnaire, and I purchased a random sample of potential customers from a database vendor.

The survey was returned with an acceptable response rate, but the results were severely underwhelming. My analysis of the data indicated the product could expect to get around 3% market share in the first year, up to 5% if you made the most optimistic assumptions.

What would you do now?

So, what would you do now? The Cheese obviously thought this was a fantastic product, so there must be something wrong with the survey pool. Yes, we had sent the survey to the wrong people! (Did I mention the random sample part?) Therefore, the decision was made to repeat the exact same survey but send it to a new group of potential customers. I dutifully followed orders, obtaining a second random sample from the database provider.

I was both intrigued and concerned about repeating the exact same survey. As a marketing researcher, it was a rare opportunity to test the validity of your survey. If the results matched, the survey plan was validated. Conversely, if The Cheese was correct and the results improved, then there was a problem with the survey, meaning you couldn't trust either result.

Second verse, same as the first

The results of Survey Two were just slightly better than those of Survey One but essentially matched the 3% market share projection. At that point, I was sure of two things. First, my survey plan and questionnaire were valid. Second, this new product was a dud that should not be introduced. Of course, The Cheese still believed the product was a winner, so unbelievably, they had me repeat the same survey a third time, hoping for a better result. (Did I mention that all three samples were random?).

Well, of course, you already know that the results of the third survey matched the first two. To add insult to injury, the results of Survey Three were the lowest of the trio. If you launched the product, expect a 3% market share, which would not cover your costs. If introduced, the product was going to be a loser.

Either way, the marketing department would be overstaffed by two people. Who are they going to keep? The hot shot they just paid a recruiter to hire away from the competition, or the poor sap who told them three times that their stellar product idea was awful? So, I bailed. (That decision didn't work out well for me.)

But there was one more try

The Cheese was still convinced they had a winner. So, after I left, they conducted a focus group, flying in potential customers from across the country to glean their opinions and buying intentions about the new snazzy composite. The focus group cost was many times more than my three surveys combined. A colleague told me that the focus group was a disaster and highly embarrassing to the company, as the participants were not receptive to buying the product at all and explicitly said so.

The main issue was that although the new product provided some benefits, doctors were highly satisfied with their current products. In healthcare, there are risks in changing from established products that work well to those from an unproven source.

Relevant Story #5

Who needs marketing research when The Cheese knows everything?

At a previous employer, I conducted market research to gauge customer attitudes and usage of our oldest and best-selling product. There was some new competition, and the information would be helpful for advertising, sales and marketing strategies, and product-line extensions.

The survey response rate was decent. I tabulated the data and issued the report listing the ten most important survey findings with the survey data.

Immediately after distributing the report, The Big Cheese called a meeting to review the findings. I found this odd in that the report had no great revelations or controversial content. I didn't believe there was anything to discuss, and in a way, I was right.

The meeting began as The Big Cheese read the first conclusion. He said, "I believe this one, therefore it is true." However, on the second one, he said, "I don't believe this one, so it's not true." He proceeded to go down the entire list in this manner. The sad thing is that there should have been discussions on the topics where his opinion contradicted the data. The product managers and Marketing Cheese sat silently, not wanting to challenge The Big Cheese. I was caught off guard, but my only defense would have been, "But the data says this." And that was already clearly presented in the report.

I could have saved lots of time and money if I had just taken the questionnaire and interviewed The Cheese on the subject. I could have written the report: This Is What The Cheese Thinks

About The Subject – And It Is Correct Because The Cheese Believes It To Be True.

It ended up with six "true" conclusions and four "false" ones. Of course, it was within those four debatable ones where many of the possible actionable items were. And those actions were never considered because those parts of the report were deemed to be false.

Relevant Story #6

Listen carefully, and then research ...

Sometimes, it pays off to stand alone – and stand your ground. Demand was expected to fall in my industry in 2006. All the experts forecasted that it would. All the models said it would. Everyone thought it would, but as the old expression goes, "If everyone is thinking the same, then no one is thinking."

And I also thought the same thing until I attended an industry conference in October 2005. One of the speakers made a short, off-hand comment during his presentation on a related topic. Everyone in the auditorium heard that comment, but only one guy understood what it meant.

The speaker's 14-word statement not only challenged conventional wisdom but also was the complete opposite. If applied to the total market, demand would rise, not fall.

My subsequent research determined that the speaker's statement was not an anomaly and that demand would climb, not fall, next year. So, I increased my industry forecast substantially. And by doing this, I stood alone. I called the leading expert in the industry on the subject and made my case. At the end of the conversation, he said, "I hear what you are saying, Don, but all indications show a significant drop in demand next year."

And, of course, since you are reading this, I was right, and everyone else was wrong. It was the best call of my career. It resulted in me getting hired by the expert mentioned above to forecast the entire commercial vehicle market eight years later, without a resume, without an interview, and being out of the industry for four years. Sometimes, it does pay to stand alone and stick to your number – and stand your ground.

CHAPTER 22

Becoming A Master Forecaster Who Stands Your Ground

Part 1 – Control & Competence

If you are in control of the process and are competent, you will be a Master Forecaster who can stand your ground. Here, we review and assemble all the pieces.

Control

From Chapter 4:

Control – You are in control of the forecasting process. You will embrace it and own it. Ownership implies responsibility, and you should take responsibility for the forecast's inputs and outputs. Your corporate life is at stake – so how many different hands do you want on the steering wheel?

Control is actually the least important of the four, meaning control of the forecasting function and ownership of the forecast. Still, because it's difficult to excel at the other three without having significant control, we will cover it first. And, be aware that I am biased here, being a "control freak". Tennis is my favorite participation sport (you control the action), and I was a pitcher when I played slow-pitch softball (you control the start of the action).

You Own the Forecast

You own the forecast - period! The numbers you put out there are yours. Yes, you have set up a collaborative forecasting process involving numerous people, but that final number is your responsibility. A major league baseball team comprises 26 players with various specialized skills and abilities attempting to win games. But only one person is responsible for the win-loss record – the manager. Often, the manager does the best job possible with underperforming players – but he still owns the record.

Our collaborative forecasting process at my final employer involved vigorous debate over the final forecast numbers. On rare occasions, I would "lose" the debate and finalize a number I disagreed with. But I personally still owned those numbers. If challenged, I would promote the assumptions that supported the forecast and vigorously defend these numbers.

When you take ownership of the forecast, there is greater motivation to produce good forecasts. The difference between doing the forecast and owning the forecast is similar to renting versus owning a house. When you own the house, you take greater care of the house and invest more in maintaining it. When you own the forecast, you take much better care of it and invest more effort in its accuracy.

If people insist on putting a target on you, then own the target. It's your job – so take responsibility for that job and perform it with excellence – whether or not you get recognized for it.

You Should Control the Process

If you are held responsible for the forecast, then you should be in control of the process. Therefore, you should build a process where you control each step. This doesn't mean you never modify the process or incorporate changes based on good suggestions or directives. But other people should not be dictating how *you* forecast.

Your corporate image diminishes when someone publicly tells you how to do your job. It implies that you are not component nor in control of your job function.

So, in response to: "What you really need to do is ...."

Your reply should be that:

1. "It is already part of the process that needs to be double-checked, revised, or confirmed."

2. "It is not part of the current process for this reason, but let me look at it again."

3. "It is not part of the current process but needs to be. Let me work on incorporating it."

All three responses communicate that you have a process and control it. And you are open to new ideas to improve it.

It is vital to have control of the forecast meetings so they do not spiral out of control or wander into areas irrelevant to the forecast discussions. If you don't, you may get stuck with a forecast number you disagree with. If you are not leading the forecast meeting, you should be the person talking the most because you should have the most to contribute.

Gaining Control If Needed

If you are currently not in control of the forecasting function, consider revamping the process based on this book. Try to secure as much control over the process as you can. You may have to do this stealthily and gradually because we all know that some people will hold onto corporate power at the expense of corporate improvement.

Not having control of the forecasting process is akin to a baseball manager not having control of the roster or batting order. If you are responsible for the results, you need to be in control of the process. Maintaining control helps you stand your ground.

Bottom line: This is your job. Take responsibility and take control.

Competent

From Chapter 4:

Competent – Your forecasts will be considered sufficiently accurate for use by your internal customers, and the accuracy should improve over time. Your boss, The Cheese, and other relevant parties will view you as proficient at your job.

This book will increase your competence as a business forecaster/analyst. To transform you from a fair forecaster into a good forecaster, a good forecaster into a great forecaster, and make a great forecaster into a Master Forecaster who can stand your ground. If you are already an elite forecaster, the book will confirm your current practices and strategies and provide some new tips and tricks to make you even better.

Establish A Process - Then Refine

Build your process based on the recommendations in this book and your company's business environment. All situations differ, so build the boat that floats the best on your water.

To improve your forecast accuracy, you will use the four stages of continuous improvement: Build a process (Plan) -> Do the process (Do) -> Measure the accuracy (Check) -> Make changes to the process (Act).

By implementing this strategy, you can continuously improve your forecast accuracy.

You build the process. Then, improve the process. Then perfect, to as high of a degree as possible. Finally, you need to trust your process. If you have built a reliable process, then it should provide reliable forecasts. The most inaccurate forecast I *"owned"* at my final employer (not counting the COVID year) resulted from not trusting our process. We thought the numbers produced by our process were *"off"* and decided to *"adjust"* them. We felt foolish nine months later when our rejected forecast proved much more accurate than our misguided adjustment.

If you don't establish a process, you will have continuous "leaks" (things you didn't account for), and your forecast accuracy will suffer.

The Statistical Factor

I will not tell you the best way to generate your numbers because that part of the process depends on your industry and company situation. Regardless, you need to know the best way to crunch those numbers and analyze your data. This may require continuous improvement in statistical methods and forecasting trends. Continuing education, training, and attending seminars are essential pieces here. Just develop all the skills necessary to be competent in your job.

Communicating Competence

Once the process improves your forecast accuracy, The Forecast Accuracy Report will communicate that competency throughout the organization. Because you are gathering more market information from various sources, including your expert panel, you will be able to answer more questions about the market and the forecast and share more knowledge in discussions.

Your internal customers should recognize the forecast accuracy has improved. When they do, tell them your new process is working well. If they haven't noticed, you should subtly remind them, referring back to the Forecast Accuracy Report. Be sure to remind your boss that the process you built is producing better results (especially before performance review time). The Cheese may even catch wind that the forecasts are better. If so, always mention the new process. This didn't happen by accident. It resulted from your planning and efforts, and you need to receive full credit for that.

The Competent Forecaster

The competent forecaster provides consistently accurate forecasts because they are based on a collection of the best market intelligence and data analysis available at the time. This information enables the forecaster to make solid, rational assumptions and develop forecast numbers based on those assumptions.

Those assumptions allow the forecaster to determine why the actual numbers differed from the forecast and to make changes in the process or the assumptions in future forecasts.

The competent forecaster does not guess or forecast based on gut feel. They rely on fact-based, well-researched data and explain this to anyone who might believe the forecast is "just a guess, because no one really knows." And in complex situations, the component forecaster may not know. However, you will still know more than anyone else at the time and make the best forecasts/predictions possible. Being competent helps you stand your ground.

Still Not Enough

Now, you have control of the process and have achieved competence. Unfortunately, as we have seen, that is not enough to stand your ground. So, we continue …

CHAPTER 23

Becoming A Master Forecaster Who Can Stand Your Ground –

Part 2 - Credibility and Confidence

As a forecaster, you must possess credibility and exude confidence to stand your ground.

Credible

From Chapter 4:

Credible – If you consistently produce accurate forecasts and keep the organization informed about the state of the industry, you will be considered credible by your co-workers and The Cheese. This is important because then you will be challenged less, especially by the company riff-raff, and you will have built up some political capital when the forecast is not within the standard.

Even if your forecasts are consistently within the accuracy standards, some people still will not believe your forecasts are credible. If people believe you're just guessing, then in their eyes, sometimes you are guessing right, and sometimes you are guessing wrong. If The Cheese believe you are guessing, then after a bad forecast, they may think they can hire somebody new to guess at the numbers and pay them less than you.

The factors that make you competent also make you credible, but credibility is more about perception. Your forecast may be accurate, but your customers and stakeholders need to trust that it is reliable, and for that to happen, they need to trust you.

People subconsciously think, "Can I trust this forecast? Why should I believe it?" whenever they see your numbers. Other skeptics will be prone to believe the numbers aren't worth much because they perceive forecasting as just guessing. So, credibility is essential, but it must be earned over time.

That is why everything in an effective forecasting process helps establish credibility. Once you become credible, people trust you. Now, we look at how the process and other tactics build credibility.

Establishing Credibility

Collaborative Team Forecasting

You are not doing "pope" forecasting, where you are sequestered in a castle and release a puff of smoke when the forecast is completed. Instead, you create a "team forecasting" environment where you bring in knowledge from people in manufacturing, sales, marketing, and wherever to help you forecast.

People will have greater buy-in of the final number when they help create it. The team approach says let's work together to manage our part of the business with excellence. Let's collaborate rather than engage in finger-pointing and bickering.

This collaboration shields you from much criticism. The VP of Manufacturing who shut down the plant manager who ripped into my forecast indeed believed he was part of the team. That action signaled to his department that it was not acceptable to criticize my forecast publicly. That type of protection is golden to a forecaster.

People who see themselves as teammates are much less likely to criticize the captain. Therefore, collaborative, team-oriented forecasting is a crucial element of establishing credibility.

The Forecast Assumptions Report

This report lists everything the forecast is based on. By supporting the forecast, it provides ample credibility. It also displays your depth of research and analysis work and your ability to obtain and synthesize all opinions and insights of other people and sources.

It indicates how hard you work to be a competent forecaster, which lends credibility to the forecast. Your assumptions will not only be highly correct, but the fact that you considered and thought about the factors adds to the credibility.

And remember, when the forecast is challenged, you will discuss the assumptions, not argue about the numbers. Arguing about the numbers, especially with The Cheese, is a losing endeavor and results in a loss of credibility. Discussing market/industry factors establishes you as an expert and builds credibility.

The Expert Panel

Building a panel of experts with whom you continuously share market/industry information builds credibility in several ways. Talking with people outside your industry, economists, for example, makes them view you as an industry expert. This facilitates future discussions, and they may even call you if they have questions about your industry or refer journalists to you.

Talking with people within the industry naturally raises your prominence. I was soon considered an "industry expert" just by talking and sharing information with other "experts". I became a member of the club by regularly communicating with other members. At that point, people began to refer others to me, strengthening my expert network. Having a solid expert network also helps if you need to find new employment.

And, of course, the expert network gains you tremendous credibility in your company. You know more than others because you talk to those who know stuff. You are going to include the best tidbits in the Forecast Assumptions Report. Use terms such as "reports from the field indicate" or "word on the street" if the information is speculative and you don't want to reveal your sources in print. It gives you an air of mystery and implies you have exclusive information. To amplify your expertise, sprinkle these '"scoops" into the conversation. Such as: "Marty over at Greystone said ...." The Cheese loves this type of stuff, and it establishes you as a credible expert.

Absolutely name-drop whenever you can. "I was just talking with Jason Fletcher over at the Cleveland Fed, and he said ...." Don't pretend it's a big deal. Act like it happens all the time. This makes you larger than life ... and highly credible.

The Forecast Accuracy Report

Publishing the Forecast Accuracy Report every month shows you are working diligently to produce accurate forecasts - ones that are credible enough that you are willing to share the results. There is a subtle implication that the forecasts must be accurate, or you wouldn't broadcast the results every month. Few people in the company would be willing to display their monthly performance. Heck, some people spend more time trying to hide their mistakes than doing actual work.

Even when you go through cold streaks, which inevitably occur, your willingness to share your results reinforces your credibility. Most people will not even read this report, regardless of whether you had a good or bad month. The production planners will usually read it monthly for clues on adapting to future forecasts.

Reasons vs. Excuses

When questioned about a forecast outside of the accuracy standards, the credible forecaster will offer reasons tied to the assumptions, if possible, but not excuses. Excuses sound petty and unprofessional and imply you are incompetent, not in control of the process, and thus not credible.

Say your forecast was too low that month, and The Cheese questions you:

Reason: "*The forecast did not account for the huge increase in Ralston sales last month.*"

Excuse: "*Sales never told me that Ralston was going to buy that much!*"

Now, if The Cheese persists, you will have to admit there was a communication breakdown, no matter who was at fault (Hey, maybe you should have known). But never throw the salesperson or sales manager under the bus. You don't like it when people throw you under that bus, so don't do it to them. You need them on your team, so treat them like teammates.

Transparency

The proposed process improves the transparency of your forecasting effort in multiple ways. You share much with Manufacturing about your shared goals and plans to achieve them. You reveal the forecast assumptions that are the basis for the forecast. You present the key factors impacting the numbers in the Forecast Meeting. You elicit differing viewpoints in the forecast meeting. You communicate the results in the Forecast Accuracy Report.

By being transparent, you gain credibility. You are not hiding or holding back anything. Well, only stuff they don't really need to see! You are showing them how the sausage is made to convince them that it is of high quality.

Communication

Opening up the lines of communication increases your and the forecast's credibility. You are getting everyone on the same page through the process, especially with the Forecast Assumptions Report. Everyone may not agree with you, but they know the factors you are using. Just the fact that you are in constant communication with key people increases your visibility and thus raises your credibility.

The increase in communication also fits into the "no surprises" strategy. If you communicate constantly and effectively, the number of bad surprises is minimized. You will always encounter people who have not read your reports and do get "surprised". But it is much easier to say, "Here it is on page 4," than "Oh, I didn't tell you."

Admitting You Are Wrong

No matter how good your process is or how hard you work, you will occasionally, hopefully rarely, be wrong. In those cases, freely admit that you were wrong. Nothing damages your credibility faster than trying to argue that you were right when it is apparent to everyone else that you were wrong.

Think about the politicians who will not admit their mistakes. The ones who argue vociferously about their decisions just after the report detailing their blunders is published. Are you likely to believe them the next time they claim something? They have lost almost all credibility, which is extremely difficult to reestablish, if at all. So, own up to your mistakes when needed.

At my final employer, people understood that the team and I worked hard to produce the best forecasts possible. But our customers appreciated it when I admitted our flawed assumptions. This gave our future forecasts greater credibility, not less, because we were not trying to portray ourselves as perfect but highly competent.

Other firms in the industry would not admit their mistakes and make flimsy excuses. Customers absolutely hated that approach, and those forecasters lost credibility. Establishing credibility enables you to stand your ground.

Confident

From Chapter 4:

Confident – Once you are competent and credible, you can confidently present, support, and defend your forecast and assumptions. This does not mean you are arrogant, although you do need to broadcast it when you hit a home run. If you are confident, you do not need to be defensive when your forecast is challenged. Instead, you can discuss it without breaking into a heated argument.

Confidence results from being in control of the forecasting process, becoming competent in your job, and gaining credibility in your company and industry. You will achieve this confidence over time until there is one moment - a presentation, a discussion, an analysis, or a spot-on forecast when you realize you have arrived. Confidence is essential if you are going to stand your ground.

Fewer Attacks

Once you gain and express confidence, most of the corporate bullies and jerk faces will leave you alone and go abuse someone else. It's no fun to punch someone who can block all your punches and will sometimes strategically punch back. You have moved the target from your back to your front so that most challenges will come straight at you, and once you've shown your ability to block shots, the number of shots will decrease.

There will still be forecasts that are out of standard and generate chatter. But you are now confident in yourself and your process. Investigate the cause as usual, but assure everyone that despite problems with "this" forecast, you have the situation under control.

Fewer Questions – Not Defensive

Fewer attacks result in fewer stupid, irrelevant, and "gotcha questions". However, you still want people to engage with you concerning your forecast. Welcome the opportunity anytime to discuss the forecast/market. You can listen carefully, without interrupting, because a confident forecaster never has to be defensive.

Don't Try to Win Arguments

First, you shouldn't have any arguments, but some people's communication style is argumentative. You have the information to argue effectively, but the confident forecaster doesn't need to argue. Present the facts as you know them; you shouldn't even have to raise your voice. Agree to disagree, if necessary. Learn, investigate, and change your current assumptions and outlooks if appropriate. This enables you to stand your ground without being confrontational.

Admit When Wrong – No Apologies

The confident forecaster can readily admit when they are wrong – because they are seldom way off the mark. It also helps when you can give valid reasons for a whiff. When I made the wrong call about the Supply Chain Crisis of 2021, this is what I said to our customers on the company webinars:

"I was wrong about the severity and length of the supply chain backup. In previous situations, the equipment manufacturers were able to find a way around the shortages and keep building vehicles; this time is much worse than anything we've seen in recent times."

So, absolutely admit mistakes but rarely apologize for an inaccurate forecast. Experts will disagree with this approach, and apologizing is recommended in most business situations. This does not apply in the forecasting realm. A confident forecaster must present themselves as strong and confident to ward off the snipers. So, unless you make a math or obvious functional error, you can admit the mistake but don't apologize for it.

Eliminate Weasel Words

Average forecasters will use "weasel words" such as: maybe, might, could, may, etc. They are acceptable, but they raise doubts about the forecast and, thus, doubts about the forecaster.

The confident forecaster will use stronger terms but still modify them somewhat because you don't know for sure, and you lose credibility if you make emphatic statements that later turn out to be wrong. So, do this:

"My research indicates that the market will increase by 10% next year."

"My models expect steady growth in the next two quarters."

"The indicators are consistent, and I see nothing that will push the industry significantly higher in the mid-term."

You will still speak and communicate with authority even in times of uncertainty, pointing out the risks and volatility. You do this because you are not guessing – you are making deductions based on your best information and assumptions at the time. But if your analysis is filled with weasel words, people may assume you are just guessing.

Overconfidence / Underconfidence

When on a hot streak, let's say five excellent forecasts in a row, there is a tendency to become overconfident. This can cause you to drop the above qualifiers and make bold statements such as: "This will happen", etc. You must avoid this trap because you are not "all-knowing," and circumstances can change suddenly.

Avoid overconfidence because it will tempt you to skip steps in your process. Remember, you are only one lousy forecast away from being shown the door.

Conversely, forecasters can be plagued by self-criticism and self-doubt. This is particularly true for you "perfectionists" out there. Even if your forecast is only 5% off, it still isn't perfect. Perfectionists make good forecasters because they are self-motivated to achieve perfection, but the psychological stress of rarely being "perfect" can grind on you.

My mood for the day changed when I measured my monthly forecasting results. If I had a lousy month, I would be dejected that day. I had lost that round. I had failed to hit my standard. If you care about your job performance, failure should hurt. However, it should not cause you to doubt yourself or your process. It should provide ample motivation to plug those gaps in your process so that future forecasts are less impacted by "leaks". You should also attempt to identify your biases and determine if you over-weighted certain assumptions.

Typically, I was only disappointed for a day. The next day, I was committed to delivering the best next forecast possible under the conditions.

If someone points out or asks about the substandard forecast, admit it wasn't up to your standards. State the reasons why it was off. And most importantly, indicate your desire to improve your process. But never, NEVER, sluff it off like it's not a big deal, especially with The Cheese. Take responsibility for the current forecast, and present yourself as a confident forecaster who fully expects to do better on the next forecast.

Knowledge Is Power

You can be confident because you may not be the smartest person in the room, but when it comes to the forecast and industry conditions, you should have the most knowledge. The trick is to dispense this knowledge in drips at the relevant times in the discussion. This way, you are not flaunting your knowledge but letting others realize you have done your research.

You do not want to come across as a know-it-all because people don't like that. Worse yet, people may believe you do know it all and ask you questions outside of your expertise.

Arrogance vs. Confidence

Once you become a confident forecaster, the challenge is to avoid being arrogant. There is a fine line between arrogance and confidence. This line will vary depending on the corporate culture. You have more leeway if The Cheese is arrogant and rewards that attitude.

On the arrogance/confidence continuum, the goal is to strike the right balance. If you are too arrogant, people will not like you. You will lose their respect, and they will be more likely to criticize you when the forecast is off. This can affect your raises and career growth. Arrogance without competence is always a loser, although we have all worked with some of those people.

At a former employer, I struck a good balance. Arrogance was rewarded there, but only amongst The Cheese. So, I exuded as much confidence as possible without crossing the line. At my final employer, I was more arrogant in my approach. I was considered an industry expert, and my image was beneficial in differentiating us from our competition. My presentation style included some swagger because I needed to present myself as an industry authority and instill confidence in my forecast.

Bottom line: If you don't display confidence in your numbers, no one else will have confidence in your numbers.

Confidence in your process produces confidence in your forecast. Confidence in your forecast enables you to stand your ground.

Two More "C" Factors

There are two more factors that we covered that are important to the master forecaster and help you stand your ground.

Communication

Throughout the forecasting process, you communicate the assumptions supporting your forecast and your forecast accuracy. You are not springing any surprises on people and are keeping all interested parties in the loop.

Creativity

The bottom-line goal is to get the correct answer. This isn't high school algebra or corporate accounting, so it doesn't matter how you get the answer; you just need it to be the right answer. Therefore, use any means necessary, within ethical limits, to get that answer. Compare and contrast. Extract data and snag information from unconventional sources. Make the best assumptions you can. And just win, baby!

The Master Forecaster

If you have: Control, Competence, Credibility, and Confidence, you are a Master Forecaster who can stand your ground.

Chapter 24

Forecast Well and Stand Your Ground

"Now, you have snatched the pebble from my hand; it is time for you to leave ...."

I wrote this book with a clear purpose: to empower you with my experience and knowledge. By applying these insights, you can transform into a Master Forecaster/Superior Business Predictor — a professional, an expert in your field, and most importantly, a confident individual.

I want you to learn from my successes and failures and build upon this book's principles to become a Master Forecaster and stand your ground. I want you to become a forecaster much better than I was because I just gave you a boost.

And if you are a young forecaster, just beginning your career, and you have read to this point, realize that I have just given you a mega-boost, and you are probably ten years ahead of your peers.

This book also aims to equip you with strategies that will help you maintain your current position and flourish in your career. These tactics are designed to make you resilient, helping you avoid the corporate axe and adapt to any situation.

Even with this, I would be remiss if I didn't tell you that following everything in this book still did not prevent me from being downsized in 2009 during the Great Recession. My industry suffered an 80% plunge from peak to trough. The Big Cheese, who never really liked me, decided they didn't need anyone to tell them sales would be depressed over the next three years. (And in this regard, they may have been correct.)

My colleagues inside and outside the company were shocked at my dismissal. Ironically, my reputation outside the company was much more esteemed than from within.

Because of that stellar reputation, I joined my final employer and worked my dream job for the last nine years of my career. Lesson: Keep working on developing yourself. You are always working for yourself – never working for a company.

The Final Word

Now get out there, use every tool and weapon I have provided, and win the forecasting battle. Become a mighty Master Forecaster and Superior Business Predictor. Forecast well and prosper! And always – STAND YOUR GROUND!

Don Ake

donake@outlook.com

www.donake1.com

www.ingramcontent.com/pod-product-compliance
Lightning Source LLC
Chambersburg PA
CBHW051529020426
42333CB00016B/1844